SCIENCE IN MEDIEVAL ISLAM

Science in Medieval Islam

AN ILLUSTRATED INTRODUCTION

by Howard R. Turner

University of Texas Press
Austin

Illustration credits begin on page 247.

Copyright © 1995 by Howard R. Turner
All rights reserved

Fifth paperback printing, 2009

Requests for permission to reproduce material from this work should be sent to:
 Permissions
 University of Texas Press
 P.O. Box 7819
 Austin, TX 78713-7819
 utpress.utexas.edu/index.php/rp-form

∞ The paper used in this book meets the minimum requirements
of ANSI/NISO Z39.48-1992 (R1997) (Permanence of Paper).

Library of Congress Cataloging-in-Publication Data

Turner, Howard R., 1918–
Science in medieval Islam : an illustrated introduction / by Howard R. Turner.—1st ed.
 p. cm.
Includes bibliographical references and index.
ISBN 978-0-292-78149-8 (pbk. : alk. paper)
 1. Science—Islamic countries—History. 2. Science, Medieval. 3. Civilization,
Medieval. 4. Civilization, Islamic. I. Title.
Q127.I742T78 1998
509'.17'671—dc21   97-7733

For Ray T. Graham,
who opened the doors

Contents

Illustrations ix
Foreword and Acknowledgments xv

Introduction 1
1. Islam as Empire 5
2. Forces and Bonds: Faith, Language, and Thought 10
3. Roots 26
4. Cosmology: The Universes of Islam 36
5. Mathematics: Native Tongue of Science 43
6. Astronomy 59
7. Astrology: Scientific Non-science 108
8. Geography 117
9. Medicine 131
10. Natural Sciences 162
11. Alchemy 189
12. Optics 195
13. The Later Years 201
14. Transmission 209
15. The New West 217
16. Epilogue 222

Islam and the World: A Summary Timeline 231
Glossary 239
Works Consulted 241
Illustration Sources 247
Index 253

Illustrations

9 *Figure 1.1.* The Early Expansion and Major Centers of Historic Islam
24 *Figure 2.1.* Nocturnal Ascent of Muhammad
25 *Figure 2.2.* Aristotle Teaching
35 *Figure 3.1.* The Genesis of Islamic Science
40 *Figure 4.1.* Diagram of Mystical Cosmos
41 *Figure 4.2.* Man and the Macrocosm
42 *Figure 4.3.* Diagram Relating to Ptolemy's Theory of Planetary Motion
51 *Figure 5.1.* Demonstration of Finger Reckoning
52 *Figure 5.2.* Development of Arabic Numerals
53 *Figure 5.3.* Demonstration of a Trinomial Equation
54 *Figure 5.4.* Proof of Euclid Postulate
55 *Figure 5.5.* Division of a Musical Chord
57 *Figure 5.6a.* Geometrical Pattern in Ceramic Tile
58 *Figure 5.6b.* Stucco Stalactite Cupola
58 *Figure 5.6c.* Ceramic Plate with Geometric Design
70 *Figure 6.1.* Constellations Little Bear, Great Bear, and the Dragon
72 *Figure 6.2.* Constellation Draco
73 *Figure 6.3.* Constellation Sagittarius
74 *Figure 6.4.* Page from Ptolemy's *Al-Majisti*
75 *Figures 6.5a* and *6.5b.* Diagrams Illustrating Epicyclic Planetary Motion
76 *Figure 6.6.* Teacher of Astronomy with Students
77 *Figure 6.7.* Ka'ba, Mecca, Saudi Arabia
78 *Figure 6.8.* Stone Sundial
79 *Figure 6.9a.* Prayer and *Qibla* Tables

80 *Figure 6.9b.* The "Mecca Plate"
81 *Figure 6.10.* Pages from a *Zij*
82 *Figure 6.11.* Ottoman *Ruznama* (Almanac)
83 *Figure 6.12.* Astronomers at Work
84 *Figure 6.13.* Underground Arc of the Great Observatory
85 *Figure 6.14a.* Samrat Yantra (Main Sundial)
86 *Figure 6.14b and 6.14c.* Rasi Valaya and Jai Prakash Structures
87 *Figure 6.15.* Astronomers Working with an Armillary Sphere
89 *Figure 6.16a.* Diagram Illustrating the Astrolabe
90 *Figure 6.16b.* Parts of an Astrolabe
91 *Figure 6.17.* Twelfth-century Persian Brass Astrolabe
92 *Figure 6.18a and 6.18b.* Eighteenth-century Persian Astrolabe
93 *Figure 6.19.* Detail of Fourteenth-century Spanish Astrolabe
94 *Figure 6.20a and 6.20b.* Universal Astrolabe
95 *Figure 6.21.* Fifteenth-century Spherical Astrolabe
96 *Figure 6.22.* Astrolabe Mater
97 *Figure 6.23.* Astrolabe, Detail Showing Date Calculator
99 *Figure 6.24a and 6.24b.* Fourteenth-century Egyptian Quadrant
100 *Figure 6.25.* Sixteenth-century Brass Quadrant
101 *Figure 6.26.* Astronomer Observing a Meteor with a Quadrant
102 *Figure 6.27.* Compass with a View of Mecca
103 *Figure 6.28.* Seventeenth-century Celestial Globe
104 *Figure 6.29.* Seventeenth-century Celestial Globe
106 *Figure 6.30.* Diagram Illustrating "Tusi Couple"
107 *Figure 6.31.* Diagram Illustrating Planetary Movement
112 *Figure 7.1. Battle between Bahram Chubina and Khusrau Parwiz*
114 *Figure 7.2.* Traditional Arab Horoscope
115 *Figure 7.3.* "Astrological Computer"
116 *Figure 7.4.* Diagrams Describing Lunar Eclipse
122 *Figure 8.1.* Astrolabist Comes to the Aid of Noah's Ark
124 *Figure 8.2.* Ship Crossing the Persian Gulf
125 *Figure 8.3.* Map of Spain and North Africa
126 *Figure 8.4.* Map of Turkestan
127 *Figure 8.5.* Map of the World
128 *Figure 8.6.* Sixteenth-century Map of the New World and West Africa

ILLUSTRATIONS

140 *Figure 9.1.* Portraits of Nine Greek Physicians
142 *Figure 9.2a.* Hospital at Divrigi, Turkey
142 *Figure 9.2b.* Hospital of Beyazit II at Edirne, Turkey
143 *Figure 9.2c.* Plan of the Hospital of Qalaoun, Cairo
144 *Figure 9.3.* Diagram of the Human Nervous System
145 *Figure 9.4.* Diagram of the Eye
146 *Figure 9.5.* Surgical Instruments
147 *Figure 9.6.* Case with Surgical Instruments
148 *Figure 9.7.* Dislocated Shoulder Being Set
149 *Figure 9.8.* Physician Treats a Blind Man
150 *Figure 9.9.* Physician and Attendant Preparing a Cataplasm
151 *Figure 9.10.* Diseased Dog Biting a Man's Leg
152 *Figure 9.11.* Boy Bitten by a Snake
153 *Figure 9.12.* Dioscorides Handing Over the Fabulous Mandragora to One of His Disciples
154 *Figure 9.13.* The Useful Chamomile
155 *Figure 9.14.* Iris and White Lily
156 *Figure 9.15.* Pharmacists Preparing Medicine from Honey
157 *Figure 9.16.* Ceramic Drug Jar
158 *Figure 9.17.* Anatomical Study of the Horse
160 *Figure 9.18.* Bath Scene
167 *Figure 10.1. A Man Gathering Plants*
168 *Figure 10.2. Small Black and White Bird on a Limb with Butterflies*
169 *Figure 10.3. Hunting Hawk*
170 *Figure 10.4. Leopard*
172 *Figure 10.5.* Selection of Fanciful and Realistic Fauna
173 *Figure 10.6.* Men Treading and Thrashing Grapes
175 *Figure 10.7a.* Waterwheel in Action
176 *Figure 10.7b.* Ninth-century Reservoir, Kairouan, Tunisia
176 *Figure 10.7c.* Khvaju Bridge, Isfahan, Iran
178 *Figure 10.8.* Farmers and Animals
179 *Figure 10.9.* Traditional Muslim Gardens and Fountains, Alhambra
180 *Figure 10.10.* Present-day Valencia Water Court Meeting
181 *Figure 10.11.* Design for a Water-raising Device
182 *Figures 10.12a and 10.12b.* Traditional Outdoor Water Clock

184	*Figure 10.13.* Design for Castle Water Clock
186	*Figure 10.14.* Basin of the Two Scribes
187	*Figure 10.15.* Design for Water Fountain of the Peacocks
188	*Figure 10.16.* Mechanical Boat with Drinking Men and Musicians
193	*Figure 11.1.* The Cosmology of Alchemy
194	*Figure 11.2.* The Philosopher's Stone
198	*Figure 12.1a.* Diagram of the Eyes and Related Nerves
199	*Figure 12.1b.* Diagram Representing Ibn al-Haytham's Theory of Vision
200	*Figure 12.2.* Diagram Illustrating Principles of the Camera Obscura
208	*Figure 13.1.* Map of Islam in the Late Eighteenth Century
214	*Figure 14.1a.* Title page from Sixteenth-century Copy of Aristotle's *De Anima*
215	*Figure 14.1b.* Page from Latin Translation of Avicenna's *Canon*
216	*Figure 14.1c.* Page from Rhazes' *Liber ad Almansorem*
228	*Figure 16.1.* Peoples of Islam Today

Maps and diagrams prepared by Michael Graham

*Under the guidance of
a series of ʿAbbasid caliphs who had a
passion for knowledge—
al-Mansur, Harun al-Rashid, al-Maʾmun—
the new civilization developed
with incredible speed and efficiency.*

GEORGE SARTON
The History of Science and the New Humanism

Foreword and Acknowledgments

This book is based substantially on research carried out in the course of my work in helping to conceive and organize "The Heritage of Islam," an exhibition of historic Islamic arts and science. Sponsored by the National Committee to Honor the Fourteenth Centennial of Islam, the exhibition traveled to five major museums in the United States during 1982 and 1983. I functioned chiefly as the curator of the scientific exhibits.

In my research and writing connected with this project, I was guided by several eminent historians of science, who helped me in every conceivable way from the beginning of planning to the installation of the exhibition. These scholars included, first of all, Professor A. I. Sabra, Professor of the History of Arabic Science, Harvard University, and Dr. Sami K. Hamarneh, Curator Emeritus, Department of the History of Science, National Museum of American History, Smithsonian Institution. Together these two served officially as the exhibition's advisors for science. The third major advisor was Professor David A. King, formerly Associate Professor of Arabic and the History of Science, New York University, now at the Institute of the History of Science, Johann Wolfgang Goethe University, Frankfurt-am-Main. In addition, valuable guidance was provided by Professor F. Jamil Ragep, formerly of the Department of the History of Science at Harvard University and now at the Department of the History of Science, University of Oklahoma, as well as by Professor George Saliba, Professor of Arabic and Islamic Science, Columbia University.

In preparing this book I have supplemented my research for the exhibition with extensive new study that the passage of time, let alone the requirements of the book, have dictated. In this connection, I am deeply indebted to Professor Michael G. Carter, formerly of the Department of Near Eastern Languages and Literatures, New York University, now with the Department of East European and Oriental Studies, Oslo University, and to Professor Ragep for their extensive reviews of my manuscript and their resulting corrections and suggestions, as well as to Dr. Hamarneh for valuable additions and corrections to the book's chapter on Islamic medicine. I am also grateful to Professor King for having provided me in recent years with much new and valuable material concerning Muslim astronomy and astronomical instrumentation, and to Professor Sabra for supplying valuable information concerning both the Hellenistic influence on Muslim philosophy and science, as well as the complex course of development and decline in the medieval Muslim scientific enterprise. I wish also to thank Dr. Alnoor Dhanani and Dr. Emory C. Bogle for their many comprehensive and constructive comments and corrections. I am, of course, entirely responsible for my interpretation and application of the valuable material, counsel, and suggestions provided by all these distinguished scholars.

This book is intended as a detailed survey for general readers and as corollary or background reading for college and high school students. While written from a Western, non-Muslim point of view, it aims to reflect full consideration of the religious and ethnic experience that has shaped the course of science in Muslim lands. Many of the illustrations in the following pages display objects included or reproduced in "The Heritage of Islam." I would like to acknowledge invaluable guidance provided between 1979 and 1982 by the following individuals and institutions in connection with my search for illustrations as well as artifacts (most of the individuals mentioned here are identified according to their positions and affiliations at the time of their assistance in the early 1980s): Richard J. Wolfe, Curator, Rare Books, Francis A. Countway Library of Medicine, Harvard Medical School; Owen Gingerich, Professor of Astronomy and the History of Science, Harvard University; and Roderick S. and Madge Webster, Curators, Antique Instrument Collection, Adler Planetarium and Astronomical Museum, Chicago. In this regard I owe a special debt of thanks to Leonard Linton, President, Central Resources Corporation, New York; his generous loan of astrolabes provided the exhibition with a uniquely spectacular asset, and he has again provided his generous assistance in connection with illustrations for this book. The following have also provided valuable insights in the selection of illustrative material: M. U. Zakariya, Arlington, Virginia; Professor Noel Swerdlow, Department of History, University of Chicago; Nina Root, Librarian,

FOREWORD & ACKNOWLEDGMENTS

American Museum of Natural History, New York; John R. Hayes and Janet Dewar, Mobil Corporation, New York; the late Sali Morgenstern, Curator, History of Medicine and Rare Books, New York Academy of Medicine; Joseph T. Rankin, former Curator, Spencer Collection, and Bernard McTigue, former Librarian, Arents Collection, New York Public Library; Dr. George Atiyeh, Head, Near East Section, African and Middle Eastern Division, Library of Congress, Washington; Dr. Esin Atil, Curator of Near Eastern Art, Freer Gallery of Art, Smithsonian Institution; Deborah Warner, Associate Curator, Department of Physical Science, National Museum of Natural History, Smithsonian Institution; Ahmad Y. Al-Hassan, Institute for the History of Arabic Science, University of Aleppo, Syria; Philip M. Teigen, Librarian, Osler Library, McGill University, Montreal; Donald Hill, Great Bookham, Surrey, England; A. Ph. Segonds, Paris; Professor Ursula Weisser, Institute for the History of Medicine, Friedrich-Alexander University, Erlangen, Germany; and Woodfin Camp and Midge Keator of Woodfin Camp and Associates.

In addition, I owe special thanks to Dr. Eleanor G. Sims, Curator of "The Heritage of Islam," who provided invaluable help in locating artifacts as well as manuscript illustrations, both for the exhibition and for this book, that I would otherwise never have known about. For essential assistance given me, apart from that already mentioned, in the course of my research for both the exhibition and this book, I want to express special gratitude to Ray T. and Roy E. Graham, Michael T. Graham, Wray Steven Graham, DeWitt Yates, A. Floyd Lattin, Geri Thomas, Stuart A. Day, and Lewis W. Bushnell, all associated with Ray Graham Associates, Washington, in the production of "The Heritage of Islam." I am particularly grateful to Michael T. Graham for preparing the fine maps and charts that will, I am sure, help greatly in orienting the reader of the pages that follow. I also owe Geri Thomas special thanks for providing crucial assistance in requisitioning illustrations for this book. Many thanks are also due the following individuals for valuable suggestions provided in recent years: Donald L. Snook; Mark Piel, Librarian, and the staff of the New York Society Library; Jenny Lawrence; Alan Pally; Isa Sabbagh; James T. Maher; John Wykert; and my friend Robert Hertzberg.

Finally, I wish to express my appreciation for the extensive cooperation and assistance given generously by University of Texas Press staff members

Dr. Ali Hossaini, Jr., Sponsoring Editor, and Zora Molitor, Rights and Permissions Manager. I also wish to express my gratitude to Lois Rankin, Manuscript Editor, Leslie Tingle, Manuscript Editor, Peter Siegenthaler, copyeditor, Sharon Casteel, Assistant Editor, and Jean Lee Cole, Designer, for their patient and meticulous work, which has contributed substantially to the book. Finally, I want to express my thanks to Elliot Linzer for his perception and care in preparing the index.

Howard R. Turner
New York, NY

SCIENCE IN MEDIEVAL ISLAM

Seeking knowledge is required of every Muslim. . . .
SAYING ATTRIBUTED BY TRADITION TO THE PROPHET MUHAMMAD

Introduction

The rise, expansion, decline, and resurgence of Islamic civilization form one of the greatest epics in world history. In the course of the last fourteen centuries, Muslim philosophers and poets, artists and scientists, princes and laborers together created a unique culture that has directly and indirectly influenced societies on every continent.

What is Islam? The word means several things. Islam is the youngest of the world's three major monotheistic religions. Islam is a way of life governing all aspects of human behavior. Islam is an intellectual and emotional force that in our time binds together about one out of every five of the world's people, uniting its adherents through a common faith and a written language, despite diverse identities of race, nation, and political affiliation. Islamic culture has always demonstrated, within a unity of spiritual vision, a spectacularly broad diversity of style and expression.

Heirs to earlier cultures of Asia, classical Greece and Rome, as well as Byzantium and Africa, Muslims took possession of their mixed heritage, preserving much of it and transforming much of it. Their cultural and political experience had a profound influence on the late medieval world of Western Europe, where Muslim achievements played an essential part in the evolution of the Renaissance and thus on the formation of later societies, including our own.

During the most recent three centuries the Western world has become familiar with many of the monuments and works of art and literature created in various Islamic periods and lands. The Taj Mahal, the great mosques of Cairo, Damascus, Istanbul, and Isfahan, the exquisite miniature paintings that enhance the historical and mythical sagas of Persian and Indian kings, the fabulous tales of the "Thousand Nights and One Night," Omar Khayyam's Rubayyat: these are but a few of the celebrated Islamic creations that we in the West now recognize as integral parts of our own cultural inheritance.

One part of the Islamic heritage has been until recent years less familiar to us, yet it has had a fundamental influence on all post-medieval lives: the historic achievement of Islamic philosopher-scientists, physicians, astronomers, mathematicians, technologists, and naturalists. Here was an elite community that included Christians and Jews as well as Muslims and comprised the first multiethnic and multinational group of its kind in the world's history. The accomplishments of this extraordinary scientific brotherhood are the subject of this illustrated introductory survey.

From the ninth century on, scientists in Islamic lands acquired, through translations into Arabic, a treasury of Greek, Indian, Persian, and Babylonian philosophic and scientific thought. They proceeded diligently to assimilate and systemize this intellectual legacy, all the while enriching it with innovation and invention, particularly in the areas of mathematics, optics, medicine, and astronomy. Their ultimate achievement was an unprecedented and harmoniously synthesized body of knowledge—the world's first truly international science.

What inspired the early scientific effort of the Islamic world? What sustained it? What obstacles confronted its progress as the centuries passed? What factors, within and beyond Islamic lands, contributed to its eclipse? What, ultimately, was the extent of the Islamic scientific enterprise? How did it influence the development of our world of science today? A look at the dynamic birth of Islamic civilization can open the way toward finding answers to these questions.

INTRODUCTION

A Note about the Gregorian and Islamic Calendars

For simplicity's sake, in the following text dates are given according to the Gregorian calendar now in use in most non-Islamic lands. However, any reader interested in deciphering dates found in historic Muslim manuscripts and on astronomical instruments may wish to relate the Muslim date to the corresponding Gregorian one. At the beginning of Islamic civilization, the Caliph 'Umar established a new calendar system, based on the first day of the year (AD 622) in which the Prophet Muhammad left Mecca. This day thus marked the start of Year One in the Islamic calendar. Since that time Muslims have preceded the date with AH (anno hegirae, representing the year of Muhammad's emigration, the Hegira), as opposed to AD (anno Domini), which has either preceded or followed the Gregorian date since being introduced in Britain in the eighteenth century. Inasmuch as the Islamic year is based on lunar months and amounts approximately to 354 days as opposed to the approximately 365 days of the Gregorian solar year, converting from one calendar to the other requires a bit of calculation. One basic equation does the job, approximately: $AD = 622 + (32/33 \times AH)$. Alternatively, $AH = (33/32) \times (AD - 622)$. The equation is derived from the fact that every 32 Gregorian years are about equal to 33 Muslim Hegira years. Thus, a Gregorian century amounts to about 103 Muslim Hegira years; 100 Muslim Hegira years equal about 97 Gregorian years. These discrepancies account for the fact that Muslim holidays and festivals may come around in any season. The Gregorian year AD 2000 will begin during the Islamic year AH 1421.

A Note about the Transliterations Used in This Book

To avoid confusion for readers unfamiliar with the Arabic language, only two of Arabic's many symbols and diacritical marks have been included in the text: the 'ayn and the 'hamza. The 'ayn signifies a sort of rough breath, the 'hamza a glottal sound. Western spelling has been used for place-names most commonly known to Western readers.

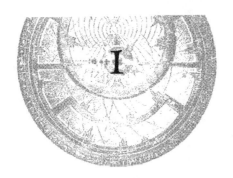

I

Islam as Empire

Within three years of the death of the Prophet Muhammad in AD 632, Arab armies, having secured control of Arabia in the Prophet's name, began advancing beyond their peninsula into territories long ruled by the Byzantine and Sasanid empires. Led by the earliest Muslim caliphs, the spiritual and political leaders who succeeded Muhammad, Muslim forces spread out explosively in all directions. They had conquered Syria, Iraq, and Jerusalem by 637, Egypt by 642, Central Asia and western North Africa by 670. Less than fifty years later the armies of Islam had invaded Spain, Persia, and India and were conducting raids across the Pyrenees. In the west their advance was finally brought to a halt in 732 near Poitiers, in what is now France, by an army under Charles Martel, king of the Franks and grandfather of Charlemagne.

Within a single century Muslims had conquered not only much of the Middle East, North Africa, and the Iberian peninsula but also parts of the Indian subcontinent. The foundation of a great empire had been established throughout lands that stretched nearly six thousand miles between the Atlantic and Indian oceans. During the fourteen centuries that followed, the outside boundaries of this empire advanced in a few areas and retreated in many others. The political empire itself split into two, three, then many caliphates, each one the equivalent of a principality. These independent domains eventually shrank, were absorbed, or disappeared. Ultimately imperial Islam came to abandon much of its political identity and almost all of its independence. But long before this decline occurred—and for nearly half a millennium—Muslim caliphs ruled over lands, peoples, and resources that rivaled in extent those controlled by imperial Rome at its own zenith seven centuries earlier.

Early Muslim conquests were accelerated by the weakened condition of the Byzantines and Persians, long afflicted with political oppression, dissension, and widespread civil disarray. Perhaps the times invited some new and compelling force, idea, or spirit. Such was forthcoming. The religious zeal of Muhammad's followers, the strengthening bonds of their new faith, their commanders' capabilities of leadership, and their soldiers' concerted military skills, superior to those of opposing tribal groups, were all crucial factors in the Muslims' conquests as they expanded both eastward and westward. The forces opposing the Muslim armies could not match the conquerors' superior strategies of attack, which derived in good part from the desert environment out of which the early armies had come and which featured the camel as basic and rapid transport.

The Arabs whose battalions spread so far and so fast belonged to a desert society long composed of farmers and nomadic shepherds, as well as a variety of merchants and commercial traders. The traditional business of this society lay in exchanging agricultural products, textiles, gold, and spices; its markets were strategically situated along the major trade routes that crisscrossed Arabia and linked it with neighboring regions on the East African coast and with India across the Arabian Sea. The rapid course of the Muslim advance across much of the Near East, North Africa, and the Iberian peninsula calls to mind the kind of carefully planned grand strategies that guided the invasion of Europe by the Allied force during the Second World War. However, the Muslims do not appear to have projected any specific agenda, let alone a timetable, for their conquest of the vast areas they ultimately won. Their initial successes impelled them to keep going. Their leaders may have heard tales of the richer lands, bigger harvests, and fabulous treasure that lay still further ahead, greater than any known in the parched environment of the Arabian heartland. Desire for the valuable fruits of conquest, however, was secondary to the driving force of what was in good part a religious and political crusade.

Muslim success in getting one vanquished population after another to accept and serve their new rulers was made easier by the long-deprived condition of so many of the people in the occupied lands and by the relatively benevolent requirements of the occupiers. Muslim rule was generally less harsh than that of previous invaders. Christians and Jews were not required to convert if they paid suitable tribute; they were also freed from obligatory (and dreaded) military service. Of course, the many sects of pantheists and pagans still risked the death penalty if they refused to pay tribute and refused conversion as well, although punishment came to be applied less rigorously in inaccessible areas.

Established on the foundation of a clearly defined but not entirely inflexible hierarchy of rulers and ruled, Islam as Empire evolved through some twelve centuries, rarely achieving political unity or equilibrium for long periods. The Prophet Muhammad, founder and head of the first Muslim state, was succeeded by four caliphs, three related to him by marriage. This group, known as the orthodox caliphate, ruled until 661, when a new and very different era began. From that point followed nearly twelve centuries of dynastic and political maneuvering and strife, including periodic war, between 1095 and 1291, with Christian crusaders.

Often overlapping in time, sometimes existing side by side, some thirty dynasties emerged, flourished, declined, and expired, their course marked by intermittent shifting of boundaries and loyalties. Between the seventh and thirteenth centuries arose the great medieval Arab dynasties: the Umayyads, with their capital at Damascus; the Abbasids, centered at Baghdad; the separate Umayyad dynasty that flourished in Spain; and the Fatimid dynasty of Egypt and northwest Africa. Together these regimes brought about the first great flourishing of Islam as Civilization. Between the eleventh and thirteenth centuries, this young civilization faced the great challenges of Christian crusaders, migrating Turks from the Eurasian steppes, and the invading Mongols from Central Asia led by Genghis Khan and his successors. These incursions by other, different societies with unique cultures of their own profoundly affected the character and evolution of Islamic society and particularly of the later Muslim dynasties that flourished after the thirteenth century, most notably the Mamluks in Egypt, the Ottomans in Turkey, the Safavids in Persia, and the Mughals in India.

Rich achievement in the arts and sciences marked virtually all of the great Muslim dynasties and imperial regimes. The expanse of Islamic culture that by the sixteenth century extended as far as Southeast Asia had become vastly diversified even as it retained its Muslim core. The evolving relationship between the culture's Islamic core and its many diverse regional ways of expression was to be Islamic civilization's most distinctive characteristic as it moved on to the threshold of modern times.

Such was the broad geographical and historical stage on which the artists, philosophers, and scientists of the Islamic world produced their work. To understand the character, extent, and quality of their scientific efforts, in particular, one should first consider the forces of intellect that inspired and maintained them.

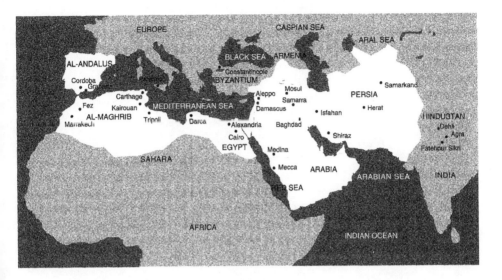

Figure 1.1
The Early Expansion and Major Centers of Historic Islam

The white area on the map indicates the extent of the Islamic domain, or caliphate, in AD 750, following its first rapid expansion. While there is not much difference between the overall territorial size of the ancient Roman and medieval Islamic domains at their greatest extent, the differences in governance were profound. At its peak all of Rome's empire was ruled by one emperor, and he could apply the system of civil law that he had inherited much as he chose. In the seventh century AD, after the reigns of the first four caliphs who succeeded Muhammad as head of state, Islam was ruled at any given time by as many caliphs as there were politically independent dynasties. But these rulers could not effectively place themselves above Islamic law, which, in good part, had been defined in Islam's holy book, the Qur'an, as well as in the teachings of the Prophet and of the orthodox schools of law. Apart from the Iberian peninsula, Sicily, and the Muslim lands of Southeast Asia, the areas dominated by Islam today are much the same as those inhabited by the Muslims at the height of the early empire, between the ninth and the eleventh centuries.

Forces and Bonds

FAITH, LANGUAGE, AND THOUGHT

Islam as Faith

Around the year AD 610, Muhammad, a successful trader and prominent citizen of the town of Mecca in Arabia, received a revelation from God while meditating in a cave. Delivered through the angel Gabriel, the heavenly message directly challenged local and traditional pantheistic beliefs with a broad range of spiritual and social concepts and commands that were, in part, parallel to tenets of both Christianity and Judaism. Thus was born a new faith, based on belief in one supreme and all-powerful God before whom all were equal and to whom all must submit. The word *Islam* itself reflects this commitment on the part of all true Muslims: it means "submission."

As a religion, Islam shares certain fundamental elements with the other two great monotheistic faiths, howsoever they vary in form and practice. Islam's holy scripture, the Qur'an, containing God's message to Muhammad, has its counterparts in Judaism's Torah and Christianity's Bible. All three religions venerate the city of Jerusalem as the site of momentous events and the location of sacred monuments marking their history. All three emphasize fundamental phenomena: revelation, final judgment, and salvation. All consider history as imbued with divine purpose.

The youngest of the three religions, Islam respects important elements in

Judaism and Christianity. The Qur'an gives Abraham, Moses, and Jesus high and honored status as earlier prophets, and it venerates Abraham as the spiritual ancestor of all monotheistic worshipers. These and other shared beliefs in no way disguise or dilute the special or unique characteristics that set Muslim doctrines apart from those of Judaism and Christianity. Muslims hold that Islam is not only a continuation of the Judeo-Christian religious heritage but also an essential and comprehensive correction of its message. Muslims honor Muhammad as the most recent addition to the roster of the great prophets; they believe him to be the final messenger of God.

The fundamental doctrine of the Islamic Revelation defines the nature of God, the role of Muhammad as God's messenger, the Qur'an as the word of God, the hierarchy and function of Islam's angels, the categories of sin, and the final day of judgment. Five acts of worship, or duties, are required of all Muslims: known as the Five Pillars, they include the profession of faith, daily prayers, almsgiving, fasting during the month of Ramadan, and, if possible, pilgrimage to Mecca once in a lifetime. Islam possesses no single, "official" church organization; in fulfilling religious obligations, each Muslim is essentially alone in the presence of God, without the mediating forms of priests and sacraments such as those found in Christianity.

Written down after Muhammad's death in the form of Islam's holy book, the Qur'an (*Al-Qur'an*, meaning "recitation" or "reading"), the Islamic Revelation represents, for devout Muslims, more than a set of religious beliefs and a system of worship. Together with the written record of actions or sayings attributed to Muhammad, known as *hadith*, the writings of the Qur'an set forth fundamental Muslim religious thought. They proclaim God's purpose: to make humankind responsible for its own actions and to place each person on earth to manage its affairs prior to finding his or her destiny in the world beyond. In providing an established form of what Muslims consider the final, perfect revelation, the Qur'an also provides a broad range of rules covering not only religious practices but also most other phases of daily life, from family relationships and personal matters of social and sexual behavior, to proper dress, eating habits, hygiene, and the conduct of business and community affairs.

Within half a century after Muhammad's death, a fierce conflict concerning the succession of Islam's religious leadership split Muslim believers into two major sects, Sunni and Shi'ite. This division has endured throughout the centuries, with Sunnis claiming about four-fifths of all Muslims living today. Shi'ites predominate only in Iran and Iraq, although they are also found in regions of Syria, Lebanon, Pakistan, Afghanistan, the Eastern Province of Saudi Arabia, and in some of the neighboring Gulf states. Over the centuries the Islamic faith has come to be practiced with variations in interpretation and ritual. Both Sunni and Shi'ite groups long ago embraced believers who adopted the doctrines and practices of Sufi mysticism. These activities are designed to help the worshiper achieve more direct and immediate communion with God through prayer, meditation, the singing or recital of devotional texts, and, within one order, ritualistic dervish dancing. The various sectarian differences have not weakened fundamental beliefs shared by all Muslims, nor did the Sunni-Shi'ite controversy hinder the rapid conquest of west and east in the century after the Prophet's death.

The desire to find a better life than was possible amid the harsh struggle for existence in the Arabian heartland, as well as the lure of the spoils of conquest, might not in themselves have inspired the early and rapid growth of the Islamic empire, nor could these motives alone have long sustained vigorous Muslim rule of such a vast territory. Islam's faith was a fundamental factor in accomplishing these epic feats. At the outset, the Muslim concept of *jihad*—better translated as "effort" or "struggle" (for the Faith) than simply as "holy war"—helped to drive Islam's forces. But *jihad*, seen as large-scale, militant struggle to advance the Faith, rarely received consistent support from Islam's leaders, especially in later centuries, when it often came to be perceived as an impractical ideal and was abandoned as an immediate option. In the Arabic-speaking world today, the term is often used to denote personal struggle to keep passion or desire under control.

What soon became the pivotal empire-building force was the generally rational, just, and humane Muslim approach to the rule and civil management of conquered territories, an approach that reflects precepts clearly set forth in the Islamic Revelation. All in all, Islamic rule encouraged cooperation on the part of local populations. Conversion became widespread in many areas, and for many people it meant less interference in their daily life than they had been accustomed to as non-Muslims or as subjects under Byzantine or Sasanid rule. Muslim law was derived from fundamental Qurʾanic doctrine, and as such tended to promote order and justice in the daily management of city, town, and village affairs. About a third of the Qurʾan's six thousand verses deal with matters of practical legislation. Within a framework of ordained universal brotherhood and individual equality, the Islamic holy book prescribes mutual help as a legal duty, proscribes useless consumption as a sin, and defines moderation in all things as an essential goal. Keeping personal pledges, exercising individual or group rights, seeking conciliation and compromise, shunning reprisal: all are demanded. Islamic law, like the faith it both reflects and defends, was from the beginning conceived to impart to each Muslim enough knowledge of his or her obligations and rights to maintain "right conduct" in this world while preparing for life in the next, thus carrying out God's will.

From the outset, within Muslim borders Christians and Jews who chose to pay the required tribute, thus retaining the right to follow their own beliefs, benefited from their special category as "people of the book." They were regarded by Muslims, respectful of the Bible and the Torah, as sharing parts of the same spiritual message received by followers of Islam and the Qurʾan. At various times some Muslims have held that the world community is split into two hostile groups, believers (Muslims) and non-believers. Of course, parochial concepts of this sort have flourished in all eras and within most great religious, ethnic, and national communities. In any event, such prejudice did not deter Muslim empire-builders from achieving spectacularly advanced levels of civilization wherever they stayed long enough, whether in Spain, in India, or in between. Furthermore, the Islamic empire at its peak embraced a population that was more ethnically and culturally diverse, as well as socially stable, than the people of earlier or later empires such as those of Rome and imperial Russia.

During the early centuries of Islam's empire, the religious and political community—the Arab word is *ummah*—was considered by Muslims to be the center of existence, recognizing its possession of God's truth and acceptance of *Shariʿa*, God's law. For several centuries notable travelers and other worldly scholars were the only Muslims to discern much of significance in the institutions and ways of life of non-Muslim peoples. However, as both commerce and conflict expanded contacts between the Islamic world and the lands beyond its boundaries, the Muslim leadership became acquainted with other types of political and social organization. Increasing contact with European societies reshaped by the Renaissance, the Reformation, and the Enlightenment led Muslims to reexamine Western constitutional and parliamentary systems of rule, which their leaders had long considered suspect and dangerous. Some of what they learned was appropriated and adapted for local use. More than any other factor, however, growing experience with imperial governance stimulated Muslim leaders to develop their own ways of applying the law to meet the increasingly complex, practical needs of communities in all parts of the Islamic realm. Their solutions to legal problems of everyday life often developed in ways that differed from those defined by religious scholars bound by holy law.

In the most traditional Muslim belief, there can be no real demarcation between religion and state or government, let alone between government and society; nor can morality be separated from politics. No earthly power has the right to invalidate or even amend Qur'anic law, let alone act outside it. The duty of rulers consists only in maintaining and enforcing it. Legal interpretation can be allowed, but it is only to be undertaken by those doctors of law or other learned individuals officially qualified for such an intricate task. The traditional view maintains that there can be but one earthly goal for all in government: justice, a society in harmony, its component parts in perfect balance within a divinely ordained system. Such a view has prevailed for centuries within orthodox Islam. However, in time, opposition between rulers and religious scholars has generated considerable flexibility involving compromise and temporary shifting of ascendancy from one group to the other.

Over the centuries, the Muslim legal picture has changed in many ways. By the end of the ninth century, the views of differing theologians had brought about the development of four schools of law, each with its own interpretation of the *Shari'a*, and the split between Sunni and Shi'ite sects generated further interpretations. Eventually, courts of appeal were created with the power to overturn *Shari'a* court judgments, and Ottoman sultans and Mughal rulers issued their own edicts to supplement the *Shari'a*. Today the *Shariʿa* remains the official foundation of law and government only in Saudi Arabia and some of the Gulf states, but its influence continues to be strong in a number of Muslim countries that have not fully adopted Western legal procedures. Moreover, the currently increasing force of Islamic fundamentalism among many Muslim communities around the world has stimulated new support for traditional *Shariʿa* concepts.

Islamic law often employs analogy in dealing with its subjects; that is, it reasons through comparison. Thought-by-analogy or association is reflected in the structure and style of Arabic speech. Whether spoken or written, the Arabic language is the second fundamental bond that, together with Islam's faith, has held the world's Muslim community together for fourteen centuries. Rarely if ever have faith, thought, and expression of feeling been served by a more appropriate and uniquely resourceful instrument.

The First Language of Islam

Arabic, the most important Semitic tongue, is today the common language of more than two hundred million people. Its alphabet is derived from the Nabataean script, which itself came from Aramaic, a language dating back as far as the fourteenth century BC. <u>Arabic is the sacred language of Islam and of the Qurʾan. In</u> its classical, or Qurʾanic form, it still serves religious, literary, artistic, and other formal needs. Classical Arabic differs in many ways from its descendant, the vernacular Arabic spoken, with considerable variation in dialect, by Muslim populations around the world.

Classical Arabic's syntax and grammar do not have the "subject . . . object" structure of Indo-European languages, nor does Arabic require the verb "to be." Most words stem from simple roots of three letters and have both specific meanings as well as universal ones drawn from their roots. The basic meaning of a root is provided by consonants (Arabic has twenty-eight of them) and is modified by changing or omitting vowels and by adding prefixes and suffixes. Short vowels appear as symbols above or below consonants. Case endings allow great variety in sentence structure. Arabic is read right to left; varying the order of words in a sentence permits a change in emphasis while retaining the basic meaning. With these characteristics, the Arabic language is an enormously fluid and flexible instrument for all kinds of written and spoken discourse, sacred or profane, philosophical or technological.

Classical Arabic has retained its original character over the years to a far greater extent than has, for example, English or French. This stability may result from its having remained for so long almost exclusively the language of religion and ritual throughout the territories dominated by Muslims. Furthermore, for at least six centuries, from the eighth through the thirteenth, Arabic was the most important language of international diplomacy and commerce. It was, at that time, also the principal language of philosophers, scientists, and poets, much as Latin was during the centuries of imperial Roman domination and as Greek was in the great age before Rome. Even today, while Muslims around the world speak in the many different tongues and dialects of vernacular Arabic, many of them read and write the same classical Arabic script, employing various Arab, Turkish, Persian, and other verbal modes and calligraphic styles. The words used by the angel Gabriel in dictating the Revelation to Muhammad—sacred Arabic words subsequently preserved for all time in the Qur'anic text—retain a living and relevant power in themselves. Indeed, devout Muslims believe the Qur'an can truly be read only in Arabic. Furthermore, reverence for the language is matched by reverence for the written script, an attachment reflected in the major role played by various styles of Arabic calligraphy in all forms of Islamic art for more than a thousand years.

Medieval Arabic's international importance was gained in good part through its adaptability: from Islam's early centuries onward, the language was stretched and augmented to serve an increasingly wide range of commercial, technological, scientific, philosophical, and literary purposes. As the basic vehicle of communication throughout the Mediterranean and Near Eastern territories, Arabic spread contemporary knowledge among people of diverse national and ethnic origins at all social levels. Quickly expanding communication between major Muslim intellectual centers in Spain and the Middle East acted as a spur to intellectual advance of every kind. Even as Muslim rule embraced more and more populations speaking different languages, from local Arabic, Persian, and Turkish dialects to those of neighboring, non-Muslim peoples, Arabic retained its central importance throughout most Islamic lands. Today it retains this power to serve the Muslim world's intellectual needs and to unite that world's peoples spiritually and emotionally.

Islamic Thought

From the start, Muslims faced the task of building a society that was not only multinational but multiethnic. Forward-looking caliphs encouraged non-confrontational contacts, especially commercial ones, with other societies. The expansion of Muslim trade helped to open Islam to outside influence, particularly from India and, to a lesser degree, China, and such traffic helped to promote advances in education and reinforced a sense of intellectual adventure. Fortunately, spiritual encouragement for such an enterprise was at hand in the *hadith*. These collected sayings, traditionally attributed to the Prophet and highly revered, include strong endorsements of learning:

> He who pursues the road of knowledge God will direct to the road of Paradise.... The brightness of a learned man compared to that of a mere worshiper is like that of the full moon compared to all the stars.... Obtain knowledge; its possessor can distinguish right from wrong; it shows the way to Heaven; it befriends us in the desert and in solitude, and when we are friendless; it is our guide to happiness; it gives us strength in misery; it is an ornament to friends, protection against enemies.... The scholar's ink is holier than the martyr's blood.... Seeking knowledge is required of every Muslim....

The assertions and recommendations contained in the *hadith* reveal the exalted position historically accorded to knowledge by the shapers of Islamic thought. Muslim religious doctrine promotes a concept of the entire material universe as a sign of God's activity, a creation by God, which God upholds. Thus, in order to understand God, it is necessary to investigate every aspect of his creation—all phenomena that exist in the world of animals, vegetables, minerals, and the elements in which people must live. This pursuit, of course, includes humankind itself as a subject. In traditional Muslim belief, such an effort to comprehend is essential in attaining the just and righteous life that forms the earthly part of a person's purpose as proclaimed by the Prophet. A simpler, more expressive spiritual motivation for scientific inquiry, let alone metaphysical investigation, is difficult to imagine.

In the early centuries of Islamic civilization the broadest possible learning was widely supported by orthodox Muslim precepts. However, in time, an opposing doctrinal trend gained strength. Not only the limitations but also the "dangers" of knowledge were increasingly described by religious authorities and pious philosophers, who declared that gaining knowledge for its own sake could never be legitimate for Muslims—it must be acquired solely as a means of gaining understanding of God, serving his will, and working to solve the problems of Islamic society. Limits to the scope of permissible learning eventually came to be defined by religious scholars, and philosophical and scientific investigation came under increasing attack as potentially destructive to faith and society. This fundamental ideological conflict has waxed and waned in other civilizations. Fortunately, during the early development of the Islamic realm, the near-obsession with the acquisition of knowledge was for a time relatively unfettered.

Another and eternal element was crucial in motivating the extraordinary intellectual expansion of the Muslim world during the early golden age of the ninth, tenth, and eleventh centuries: simple human curiosity. This drive, together with the new civilization's faith, its Qur'anic teachings about the world, and its increasingly resourceful language, impelled Muslims to set about investigating everything around them. They were equipped to make remarkable sense of their world, and this they did, with an enthusiasm and energy rivaled only in the heydays of Renaissance Italy, the Scientific Revolution, the Age of Enlightenment, and the early period of the Industrial Revolution. The cultural result of this effort was soon evident, and it amounted to notable and unique accomplishments in both the arts and sciences. In these pages, a broad survey of the Islamic scientific enterprise, in particular, can best begin by looking at the intellectual landscape in which medieval Muslim scientists undertook their work.

Islam was born without what the Western world recognizes as "philosophy." However, from the very beginning Muslims, like all others in all times and places, sought answers to questions about every aspect of existence, about God, the creation of the universe, the destiny of humankind, and the proper organization of individual and community life on earth. For traditional Muslims, the Qur'an—the holy text—has always embraced the basic answers to these questions, answers that are beyond dispute, if not always beyond interpretation. Qur'anic statements are subjected by orthodox Muslims to *kalam,* a theological discipline involving rationalist dialectical examination, parallel in some ways to the kinds of argument that have been associated with Western theological and philosophical debate since the beginning of intellectual speculation in classical antiquity. From the earliest days Islamic scholars showed great zeal in uncovering and exploring in depth the intellectual heritage they came upon in the newly conquered non-Muslim regions. In those lands, first of all, lay a treasury of philosophical writings setting forth older systems of thought, notably those of ancient Greece, India, and Persia.

The rapid expansion of imperial Muslim rule was followed by the conversion to Islam of many Christians, Jews, and others. Arab Muslims were thus exposed to ways in which thinkers holding to other faiths and ideologies approached the fundamental questions of life and being. The early Muslim philosophers came to see that their own thinking could be enhanced—or, at least, made more orderly and productive—if they adapted to their own purposes some of the speculative processes and systems of philosophical and physical classification established centuries earlier.

The works of Pythagoras, Socrates, Plato, Aristotle, the Stoics, and Epicurus appealed to Muslim philosophers. Here was a rich intellectual accumulation that they felt they could and should put to Islamic use. By the time the Islamic empire had been established, this legacy amounted to an unusually eclectic blend of philosophical concepts and argument. Ready at hand were the Socratic belief in knowledge as good and as within reach of humankind, as well as Plato's emphasis on the geometric structure of the material world and the ultimate authority of divine principles over natural laws. Here also were Aristotle's views of causality, his belief in design and purpose as the focus of all natural process, his original and comprehensive system of logic, applicable to all thought, and his macroscopically to microscopically arranged methods of classifying everything in the material universe. To be sure, this treasury of thought had undergone some evolution: major Platonic and Aristotelian concepts had become somewhat blended with mystical elements set forth by the third-century philosopher Plotinus, as well as with features of Persian, Egyptian, Jewish, and Christian belief. Ultimately called Neoplatonism, this extraordinary hybrid philosophy profoundly influenced not only most of Islam's philosophers but also many medieval Christian theologians, attracting them in varying degrees toward an overall detachment from the everyday or "real" world. During the Islamic civilization's early centuries this newfound classical and Eastern treasury of knowledge often appeared to orthodox believers as "wisdom mixed with unbelief."

Traditionalists could only view with alarm the gradual dilution of faith with rational speculation. Charges of heresy came to be leveled at such Muslim groups as the Muʿtazila, a ninth-century school of thinkers who wanted to apply reason to matters of faith and who held some clearly unorthodox views concerning the creation of the Qurʾan, free will, and predestination. The Muʿtazilites eventually fell into strong disfavor and were suppressed. As for those the Arabs called *falasifa*, philosophers who were loyal, in particular, to the principles contained in a Hellenistic version of Aristotelian thought as a whole, their view of God's nature and role and of the world's creation was not easily nor substantially accommodated by orthodox Islam. In sum, their philosophy rested on speculation dependent on human faculties and thus directly challenged the supremacy of divine Revelation.

During Islamic civilization's first few centuries, the Muslim attachment to philosophical speculation was strong and adventurous enough to survive periodic attempts to purge concepts that were less than purely Islamic. Elements of Hellenistic philosophy soon penetrated Muslim thought deeply and permanently. In the long run, even amid intermittent religious-philosophical confrontation and conflict, philosophy in medieval Islam remained sufficiently vigorous to produce a number of outstanding giants who rank among history's great intellectual movers and shakers. Five of these stand out, in particular. One way or another, all provoked controversy, and most earned strong orthodox opposition.

First and greatest of the Arab philosophers, al-Kindi, who was active in ninth-century Baghdad, developed a system of thought that encompassed religion, politics, and the sciences, achieving something of a bridge across the chasm between belief and reason. One of the first to sponsor translation of the works of Aristotle, al-Kindi was also one of the earliest Muslim polymaths, gifted in physics, or "natural science," mathematics, optics, music, and cosmology, as well as philosophy. Islamic civilization produced many such polymaths, much as a later age produced Leonardo da Vinci and the other artist-scientists known as Renaissance men. Al-Kindi was Islam's first major music theorist, and his universal interests also included meteorology. His deep involvement in the workings of the physical, or "real," world around him was to remain a significant element in mainstream Islamic philosophy throughout its period of greatest vigor. Convinced of the value of rational thought, al-Kindi was one of the first noted Muslim philosophers to become suspect in the eyes of rigidly orthodox believers.

Known as the "second teacher" (after Aristotle), Abu Nasr al-Farabi, a Turk who died in the tenth century, was particularly interested in reconciling Aristotelian and Platonic ideas. He found areas of accord between classical Hellenistic philosophy and prophecy. Accepting the validity of both the Qur'an and the philosophy of Plato and Aristotle, he reasoned that both must agree, and he worked to clarify the reconciliation. His work *Al-Madina al-fadila* (The Good City) set forth ways of gaining happiness through politics and showed the relation between Plato's ideal community and Islam's divine law.

Like al-Kindi, al-Farabi defended rational thinking. In his *Kitab ihsa' al-'ulum* (Catalog of Sciences), after listing and describing the various sciences, he placed philosophy at their head, claiming that it guarantees the truth of knowledge gained by clearly proven reasoning. Al-Farabi and al-Kindi did much to set the main course of Islam's philosophy, and the great majority of great Muslim philosophers and scientists who followed them were, essentially, their intellectual descendants. One of the most conspicuous differences between science in the medieval Islamic world and twentieth-century science in the Western world is the prominent role of philosophy in providing most medieval Muslim scientists with an almost tangible framework for their work.

The eleventh-century Persian philosophical and medical genius Ibn Sina, known in the Western world by the Latin name Avicenna, dealt with most of the eternal questions posed by rational philosophy as well as by religious faith. In dealing with the eternally challenging nature of universals—fundamental qualities, characteristics, or properties of things—he tried to reconcile Platonic with Aristotelian concepts. At times his unorthodox beliefs, along with his worldly lifestyle and elitist arrogance, aroused controversy—after his death his books were burned by an Abbasid caliph. However, Ibn Sina's all-embracing vision took effect far beyond Islam, ultimately influencing the evolution of the scholastic philosophy that dominated the medieval theological philosophy of the Christian West.

The eleventh-century jurist-theologian-philosopher al-Ghazali played a pivotal role in reinvigorating Islamic thought. His two greatest texts dealt with what he termed a revival of the religious sciences and the incoherence of the philosophers. He placed the mystical thinkers, the Sufis, above the philosophers in their ability to reach the truth. Yet, like the dialectical theologians, he accepted the role of reason in explaining divine concepts. As a result of the work of al-Ghazali, the Muslim world became more receptive to Sufism, and Islamic thought became more inclusive, more harmonious, better able to integrate philosophical, theological, and mystical elements. His work greatly influenced subsequent Christian and Jewish theologians and philosophers.

Finally, Ibn Rushd, also known as Averroës, was the brightest light of the cultural zenith that was achieved in Muslim Spain during the eleventh and twelfth centuries. He became the most celebrated of all commentators on Aristotle. Like Aristotle and some Muslim dialectical theologians, he held God's existence to be provable on rational grounds alone. Ibn Rushd came to be regarded by some as the father of free thought or even unbelief, yet his influence was not hindered by this prejudicial label, and he left a mark on both Christian and Jewish thinking.

In the development of Islamic philosophy, both within and beyond Qur'anic boundaries, there can be seen a sort of double standard concerning the dissemination of philosophical wisdom to the public. This is particularly evident in the thinking of both al-Kindi and Ibn Rushd: these two concluded that the educated elite could be guided by reason, the less-advantaged masses only by faith. More specifically, the Qur'an could be interpreted allegorically for the intellectual minority but should be delivered more as literal truth for the majority, whose faith might be undermined by "unsuitable" speculation and doubt. An unequal distribution of doctrine such as this is, of course, also characteristic of some Christian practice. In any event, as it evolved, Islamic thought endeavored to accommodate elements of worldly philosophy while retaining the directness and simplicity of the Faith as first proclaimed. Striving to have it both ways undoubtedly has as long a history as does philosophy itself, and the result never seems to include a long-lasting resolution. In the case of medieval Islam, this often complex and sophisticated endeavor was destined to continue through both fruitful and lean times into the modern age.

Although Muslim scholars gave close attention to the manuscripts of classical philosophy that they found stored in their newly won lands, what first attracted their interest were works devoted to medicine and astronomy, in which more than one of the greatest Greek philosophers had excelled. These texts offered vital guidance in practical matters given high priority by the Muslims: health, travel, and precise orientation in time and space in connection with religious practice. Notwithstanding the Muslim concern for preparing for life in the hereafter, Muslim scientific effort, from the very beginning, focused most immediately on gaining knowledge that could be applied in making life better and more efficient on Earth. This effort seems, in the light of history, to have been perfectly timed. In the eighth century most of the civilized Mediterranean and European world languished in generally low estate. Today no serious historian dealing with that world refers to the dozen-odd centuries of medieval times as simply constituting the Dark Ages, but it remains reasonable to claim that the first few of those centuries saw little of the light of civilization. Soon after the Muslims had reached as far as Spain and India some illumination began to be restored. At first much of it was the result of ancient fires being relit, but new light was soon to come as well, all of it generated by the vigorous and enterprising civilization that had begun in the desert communities far away on the Arabian peninsula.

Figure 2.1
Nocturnal Ascent of Muhammad, MS Illustration from
Guy U Chagwan (The Ball and the Stick), by 'Arifi,
Persia, Fifteenth Century

Prior to the establishment of Islam, Muhammad is believed to have been summoned by the angel Gabriel to mount a winged creature named Buraq, on which the two were carried through the skies from Mecca to Jerusalem. After praying there with Hebrew and Christian prophets, he was carried by Gabriel up through seven heavens representing ascending levels of spiritual reality, finally ascending to the highest limit, where he beheld the signs of God. In this presence, he received God's commands as to obligatory worship for all Muslims. On returning to Mecca, Muhammad spoke of his experience only to his immediate companions. This epic event, however, spread through tradition and became central to orthodox Muslim concepts of the cosmos. It may also be interpreted as pointing symbolically to the divine source of all knowledge. Beautiful illustrations of the Night Journey are found in manuscripts from Persia, in particular. The Prophet is often displayed with an obscuring veil or without facial features.

Figure 2.2
Aristotle Teaching, MS Illustration from *Manafiʿ al-Hayawan*
(On the Identification and Properties of Animal Organs),
compiled by Ibn Bukhtishu, Thirteenth Century

As important to Islam as to the West, the influence of Aristotle's writings on medieval Muslim philosophers and scientists is difficult to overestimate. His thought, along with that of Pythagoras, Socrates, Plato, and much of the rest of the classical Greek pantheon of thinkers, penetrated virtually every scientific discipline throughout the Islamic world, providing mathematicians, astronomers, and practitioners in all the physical and natural sciences with new insights into the investigation and classification of both ideas and material objects. Aristotle's position as the Great Teacher is reflected visually in many Arabic manuscripts such as this one, a translated collection of nineteen of his essays detailing the medicinal properties of animal organs. Here the philosopher, at right, lectures a student, the text in question open on a stand between the two. The halos around the heads have no hierarchical meaning and serve simply as visual emphasis. The darkness of Aristotle's face is probably caused by degeneration of the pigment used by the illustrator.

Roots

The family tree of science traces a complicated ancestry. Egyptian and Babylonian processes of scientific inquiry that developed three thousand years before the birth of Christ were the precursors of Hellenic or Greek investigations, which in turn produced Hellenistic and Harranian (pre-Islamic northern Mesopotamian) and, in part, Persian science. All these influences nurtured the later scientific enterprise of Islamic civilization. This network of transmission was further extended before and during the Islamic era by direct connections, often commercial, between Egypt and other parts of the Hellenistic world, by links between Hellenic and Harranian science, and, last but not least, by important influences from India and China, the first passing through Persia, the second directly, if somewhat intermittently, carried by travelers to Islamic lands.

These varied legacies produced a far more international, intercultural, and abundant inheritance of intellectual riches than had ever before been received by any single civilization. It was a treasury long forgotten or unknown in much of the medieval world up to the Muslim conquests in the seventh century AD. Most important to it were the manuscript remnants that contained the greatest achievements of Greek science, those of the fourth and third centuries BC, in particular, as well as the achievements of Hellenistic scientists five centuries later. These works, long ignored, were found from the eighth century on by Muslim leaders and scholars when they explored Byzantine libraries and other collections of rarities.

The disintegration of Roman civilization during the early centuries of the Christian era was marked by the general deterioration of most of the basic endeavors that enable society to survive and flourish, from producing food in abundance to generating original ways of improving the conditions of daily life. These were not centuries of much significant or extensive cultural creativity. In Europe in the early Middle Ages intellectual and social progress was fitful when not atrophied. The times were more than usually propitious for superstition, belief in magic, and reliance on every kind of prophecy or divination—whatever might help to shore up faltering faith or provide a gratifying substitute for undeniable logic.

The Muslim conquerors at first gave little indication that they would revitalize cultural life in the communities they came to dominate politically and economically. The Arabs' pre-Islamic civilization had not been marked by notable scientific advance. Members of wandering Bedouin tribes or traders, Arabs and Berbers inevitably gained considerable practical knowledge of geology, plants, and animals in managing to survive in the inhospitable landscapes of the Arabian peninsula and North Africa. They knew well the limited pasture and the oases, caravan trails, and waterways in and adjacent to their lands; they had learned to use the stars in time-telling and travel. But the emergence of Islam, the vigor of the Muslim conquest, and the accompanying enthusiasm for exploring and exploiting the intellectual and commercial riches waiting in every major community that they conquered soon propelled Muslims beyond the development of practical technology. The time was indeed at hand for the extraordinary intellectual, artistic, and scientific progress achieved throughout Islamic lands in the centuries between 800 and 1600.

The development of the Muslims' imperial civilization turned cities such as Damascus, Baghdad, Cairo, and Córdoba into great cultural and commercial capitals, kept in regular contact through expanding networks of land and sea routes. Sponsored largely by caliphs and their royal courts, architecture and the decorative arts flourished, each in ways uniquely regional, yet each in ways recognizably Islamic. Great universities were founded, each one serving thousands of students. By the twelfth century the civilization of Islam was sophisticated far beyond that of Christian

Europe. Its society was increasingly urban in character, and its thriving commerce, dominating the entire Mediterranean basin, reflected shrewd, practical, and productive management of people, goods, and wealth. Scholars in the West have long given most of their studies of medieval history a Eurocentric orientation, despite the fact that for at least six centuries, Islam was a uniquely vital cultural and commercial force.

In diverse national and ethnic ways Islam's culture displayed a unified religious spirit long after Islamic imperial power began to wane at the end of the tenth century. However, that spirit was increasingly challenged by a spreading worldliness, a growing focus on earthly as opposed to "celestial" or spiritual matters. This confrontation affected all areas of Muslim life, including the sciences. We must remember that we are speaking generally of science written and carried on in the Arabic language, rather than simply of Muslim science: from the very beginning the Islamic scientific community included many Christians and Jews as well as Muslims, and it embraced Indians and Persians as well as Arabs. This ethnic and religious mixture probably rendered early Muslim scientific endeavor more resistant to theological constraint than it would have been had only Muslim Arabs been involved.

In any event, ancient Egyptian, Mesopotamian, Persian, and Hindu achievements greatly influenced virtually all scientific development during the first few centuries of the Islamic empire. What seems remarkable is the fervor with which Muslim scientists as a whole embraced the diverse heritage they discovered in their new domains. As a crusading faith, Islam could have been expected to try to eradicate or denigrate most achievements of what were often ideologically perceived as profane, misguided, or inferior societies of the past. This did not happen; instead, the ancient legacy was treasured and utilized almost from the time of the conquerors' arrival. As a first step, an efficient and large-scale means of transmission from past to present was rapidly organized and put into service.

The Period of Translation

The city of Gondeshapur in southwestern Persia was captured by the Arabs in AD 638; it became a center for the dissemination of Greek and other ancient philosophical and scientific knowledge throughout the new

Islamic empire. Here flourished a large community of Nestorians, members of a Christian sect that had been charged with heresy and forced to flee from Christian territory late in the fifth century. The city had long supported a notable academy of scholars and physicians, many of whom spoke Greek and Sanskrit as well as Syriac, an Aramaic dialect then in wide use throughout the region. After the Arab conquest, these scholars quickly became familiar with Arabic. An intensive program of translation into Arabic of philosophical, medical, and other scientific manuscripts was sponsored by the new Muslim caliphs, and soon this rapidly expanding enterprise spread to Baghdad and Damascus, the cultural centers of the Abbasid dynasty.

Without historical precedent, the quantity of translations produced in Gondeshapur represented from the start an ecumenical as well as an international effort, involving many Christians and Jews as well as Muslims. It was most often an enterprise prompted by royal decree, reflecting interest on the part of caliphs and other high officials of court and government not only in practical sciences such as medicine and astronomy but also in the less exact precepts of astrology and alchemy. These two exotic disciplines, of course, had already enjoyed centuries of official and public support and would continue to do so for centuries more, in both East and West.

A parallel translation program, from Greek to Latin, was underway at the time in medieval Christian Europe. Sponsored by monasteries such as Monte Cassino and by rulers such as Charlemagne, its goal was the same as that of the effort being undertaken in the Islamic lands: to produce translations of classical Greek texts. This effort, however, did not match the Muslim output in scope or quantity until the twelfth century, when imperial Islam had begun to lose some of its political and cultural supremacy.

Two centuries of Muslim translation succeeded in making the major works of Plato, Aristotle, Euclid, Archimedes, Hippocrates, Galen, Ptolemy, and many others available to Muslim scholars in settlements from Persia to Spain. Not all the translations were of the greatest fidelity or quality, but revisions became frequent as scholarship matured. Flexible and capable of accommodating new concepts, procedures, and the details of science as well as of philosophy, the Arabic language proved to be a resourceful vehicle. At the outset of this enterprise emerged outstanding translators, such as Hunayn ibn Ishaq, a ninth-century Nestorian Christian who was expert in four languages and who produced a mass of work covering much of Greek philosophy and medicine. Another eminent scientist, Thabit ibn Qurra, was particularly noted for his translations and revisions of works on logic, mathematics, and astronomy. A member of a family of distinguished scientists and scholars, he never converted to Islam; he was a member of a pagan sect, the Sabians, who were mentioned in the Qur'an and thus had protected status.

In the early selections of works to be translated, priority was given to subjects deemed most immediately useful, such as medicine, mathematics, and astronomy, the latter two needed in orientation for religious practice and for astrological prediction. However, within two centuries an encyclopedic classical treasury was available in Arabic, embracing Greek knowledge and Hellenistic contributions. This store of cultural riches was capable of filling the intellectual needs of a new civilization even as that civilization was pursuing spiritual, political, and cultural aims radically different from those of ancient Greece and Rome. By the eleventh century Muslim rulers had established appropriate institutions designed to preserve and maintain this treasury and put it to Islamic use.

Islam's first few centuries witnessed the establishment of great libraries and centers of learning in both East and West. The most celebrated of these were located in the seats of political or imperial power. A famous "house of wisdom" (*Bayt al-hikma*) flourished in ninth-century Abbasid Baghdad; it was a center not only for translation but also for specialized scholarly study and investigation. Notable libraries were established in several Persian cities. In tenth-century Cairo, under Fatimid dynasty rule, one of the libraries contained forty rooms holding many thousands of works devoted to the ancient sciences. In Umayyad Spain, the great capital of Córdoba was, like Cairo and Baghdad, a center of higher learning that drew students and visiting scholars from all regions of Islam. Cairo's al-Azhar, founded late in the tenth century as a training school for preachers, later contained mosque schools (*madrasa*) and still flourishes today. After nearly a millennium of uninterrupted existence, it is the world's oldest university or college.

A science academy was established during the tenth century at Córdoba, and another at Toledo. Generally, however, advanced education in science was carried out on a personal rather than an institutional scale, within the confines of royal courts and royally sponsored establishments such as observatories and hospitals. A young man seeking high-level education in the sciences had to wait until he had progressed beyond elementary school and the *madrasa*, where he could receive instruction in basic mathematics. University curricula might include advanced mathematics, some astronomy, medicine, and natural sciences such as biology, but these subjects were treated largely within a framework of Qur'anic fundamentals and interpretation as well as principles of religious law. At the caliph's court, however, the young man might be fortunate enough to be admitted to a circle centered on one of the great teachers—often a person celebrated for achievement in more than one science, a philosopher-physician, perhaps, or a mathematician-astronomer.

Such high-level scholarship represented the upper track of a two-track system. On the lower track, those of the general population who went through elementary education were offered mostly the curricula and concepts that were officially deemed appropriate or adequate for the many. In addition, the procedures of traditional Muslim elementary education did not promote original speculation. Listening, reading, recitation, and preparation by rote formed a great part of the process of learning. As for prospects for young women who might be interested in learning more about the sciences or any other major academic subjects, they did not exist to any significant degree in any Islamic community until the second half of our own century.

The character of Islam's specialized scientific education can be better understood if one considers the traditional Islamic concept of knowledge and the ways in which knowledge and the curriculum were organized and classified in early Islam. Not only was the pursuit of knowledge exhorted by many *hadith*, the statements traditionally attributed to Muhammad, the Qur'an emphasizes the value of knowledge in grasping the nature of the world around us. As previously noted, Muslim religious doctrine defines the universe as a sign of God's activity; therefore, study of that activity is thought to provide knowledge of the right path toward the proper life on earth and salvation in the life beyond. Muslim religious leaders and educators devoted considerable time to classifying the sciences and clarifying their specific functions as they would help each person to acquire knowledge within an intellectual framework that would serve God's purpose.

Definitive classifications were attempted by such intellectual giants as al-Farabi, Ibn Sina, and Ibn Khaldun, the greatest of the Muslim historian-philosophers. Ibn Khaldun's system reveals basic distinctions that were widely adopted. First in his schema came the traditional religious sciences, dealing with examination and readings of the Qur'an, the science of tradition, sciences of jurisprudence and dialectics (argument), speculative theology, mysticism, the interpretation of dreams, and philological science.

A second major group, the philosophical sciences, included, first, Aristotelian logic; second, in a sort of umbrella category, physics, including medicine, agriculture, magic, alchemy, and other esoteric disciplines; third, metaphysics; and fourth, mathematics, covering numerical sciences such as arithmetic, calculation, algebra, commercial transactions, the apportioning of inheritances, geometrical sciences such as spherical and conical geometry, surveying, optics, and astronomy, including tabulation and astrology. Finally came music, which was considered a mathematical discipline.

Whatever the variations in degree, the different systems for the classification of learning all reflected a fundamental acceptance of the supremacy of the Islamic Revelation. Eventually there developed profound suspicion, even fear, of influences that might be exercised by intellectual disciplines inherited from the ancient world. There evolved a clear distinction between "revealed" subjects connected with religion and centered around the Qur'an and those not connected with religion, considered "rational" subjects. What today we would call physics, biology, agronomy, and the earth sciences belonged to this latter group. As a whole these sciences, termed "philosophical" because they were considered accessible through human reason rather than through faith, were regarded as foreign because of their origins in classical Greece and India. Among them, philosophy in particular, at first praised as a valuable tool inherited from the Greeks, was singled out by some orthodox scholars and leaders as alien and suspect. Ironically, the eleventh-century theologian al-Ghazali's belief in the value of logic in dealing with Islamic legal subjects helped to spread the acceptance of rational process as a valid tool in organizing knowledge.

The mosque schools and the universities' legally oriented faculties took more or less exclusive charge of teaching the religious disciplines, leaving the philosophical, mathematical, and natural sciences to be taught mostly in the small, royal, elite groups described previously. There was little of a democratic character in medieval higher education, within or beyond Islamic lands.

A dynamic new faith, the vitality of a people freeing themselves from a struggle against the harsh elements of desert life, a pressing curiosity about the world that they encountered outside the Arabian heartland, a need for a share in that world's bounty, a language well-equipped for intellectual exploration: all these factors, binding together Muslims in greater unity than existed among any other civilization at the time, generated a powerful motive for Islam's scientific enterprise. The world, all its forms of life, and the heavens above were now subject to an unprecedented exploration and examination by Muslims and other scholars in Muslim lands. And even where spiritual motivation was less pressing than material enthusiasm and practical curiosity, there was power and material reward to be gained from learning more about what the world was made of and how such knowledge could be put to beneficial and profitable use.

In less than four centuries after the first Islamic conquest, philosophers, mathematicians, botanists, physicians, geographers, alchemists, and their peers in the other scientific disciplines, at work throughout the wide reaches of the Islamic empire, had accomplished the remarkable feat of unwrapping the vast intellectual legacy received from past civilizations, analyzing it in relentless detail, fussing with it, testing and re-testing its hypotheses and answers, evolving new ones, and further revising, discarding, or reformulating many of those. From the broadest concepts of the physical universe to details of the smallest scale—including invisible processes within the human body—much was put in order and connected in ways some of which appear to reflect or to parallel the Muslim concept of cosmic unity spelled out in the Islamic Revelation. In sum, a new concept of the universe was put together, in many ways remarkably similar, or at least parallel, to the old ones, in other ways significantly simplified and clarified. More important, from macrocosmic to microcosmic element, this Islamic universe appeared, in general, orderly, functional, *workable*. It made sense. It represented a critical advance beyond contemporary non-Muslim concepts, most of which it would eventually affect profoundly.

What, specifically, did Islam's medieval scientists accomplish? It seems appropriate to begin the inventory with Muslim ideas about the overall cosmos and all that was in it.

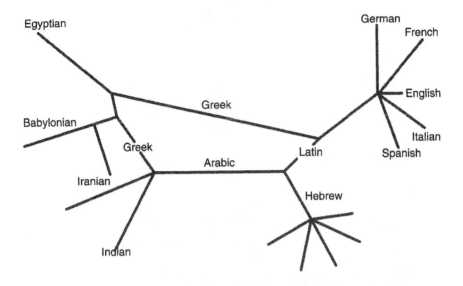

Figure 3.1
Diagram of the Genesis of Islamic Science
(The Transmission of Science from Antiquity to the Middle Ages)

Nearly half a century ago George Sarton, the celebrated historian of science, used a diagram similar to this one to show how the Arabic scientific effort revived and continued not only Greek science but also many scientific concepts of Iranian and Hindu origin. As can be seen here, the transmission of historic scientific traditions never takes a single, direct route. Alternate paths and crosscurrents were involved in all the great eras of early scientific progress.

4

Cosmology
THE UNIVERSES OF ISLAM

Cosmology, the study of the history, structure, and workings of the universe as a whole, has evolved through the millennia in several forms: mythological and religious, mystical and philosophical, astronomical. The ancient Babylonians and Egyptians, constructing their systems from a jumbled mixture of ancient myths, believed that the cosmos was a box, with earth at the bottom. Mountains at the earth's corners held up the sky above. The Nile, flowing through the earth's middle, was a branch of a greater river that flowed around the periphery. On this river traveled the sun god's boat, making its daily journey. A later, Mesopotamian concept treated the universe as a hemispherical vault that contained the flat disk of earth surrounded by waters. Waters also formed the heavens above the vault; here dwelled the gods—the sun and other celestial bodies. They appeared every day and governed everything happening on earth. Their regular orbits through the heavens were believed to control human destiny.

By the seventh century BC, Greek philosophers were proposing increasingly specific details, some of which seem almost to anticipate modern cosmological theory and findings. Thales' heavenly bodies were solid objects; Anaximander's living creatures arose from moisture evaporated by the sun; man, in the beginning, was something like a fish; the creation of the world was brought about in a series of what appear to have been explosions of steam. Philosophers such as Pythagoras and Plato viewed the heavenly bodies as godlike beings moving in absolutely uniform and circular paths. Anaxagoras denied the divinity of these bodies but believed their motions to be governed by a soul or mind.

During Greece's golden age cosmic concepts became mathematical, using geometric forms to represent the four elements—fire, air, water, and earth, as well as quintessence, the material of the heavens, with a sphere serving as the overall universe. Pondering the various mythical and physical phenomena specified by his predecessors as representing the birth, development, and management of the cosmos, Aristotle classified those he could accept into a concretely reasonable but rigid system of cosmic mechanics. He conceived of the cosmos as a system of concentric shells containing the heavenly bodies. These spheres were actual physical bodies, arranged concentrically and rotating, one within another, each sphere transmitting its movement to the next one below. The movements of the seven planets were transmitted through the highest sphere by the Unmoved Mover, related to the sphere as a soul is to a body. As a whole, classical Greek cosmology was imbued with a firm belief in basic laws of order and harmony.

The Chinese, meanwhile, had already developed their own versions of the cosmos. The Taoists of six to four centuries before Christ defined and described the two principles, yin and yang, female and male forces, passive and active, produced by matter and energy and responsible for maintaining the universe through interaction. One Chinese concept of cosmic structure involved a hemispherical vault (the sky) under which rested a convex square (the earth). Later came the theory of the celestial sphere, a spheroid universe; still later was proposed a theory of empty and infinite space, without shape or matter, in which winds moved the heavenly bodies. Early Chinese cosmology, like that of some of the ancients further west, spelled out phenomena that seem to prefigure astrophysical notions held in our own time, such as primordial matter spiraling through space and cosmic winds "blowing" out in vast streams from the sun.

Early Christians, somewhat like the Near East ancients before them, pictured a flat earth sandwiched between subterranean and heavenly waters. Meanwhile, the idea of concentric spherical shells containing seven moving planets gained popularity: its Platonic and Aristotelian character was refined by the Hellenistic astronomer Ptolemy. Many early Neoplatonic and Christian cosmological concepts featured angelic beings responsible for moving the planets on their courses within these shells. Such divine dynamos would remain as cosmic icons for centuries. However, by the time Islamic civilization was fully established, Muslims were beginning to develop cosmological schemes that were sufficiently complex and sophisticated to include as empirical fact celestial happenings that could actually be observed, such as details of the variations in planetary paths.

In early medieval Christian Europe, virtually all intellectual activity directed at understanding the creation, form, and management of the cosmos drew primarily on religious faith or superstition. Concepts founded on reason alone risked being condemned by the Church as heretical. However, in investigating the nature of the cosmos, the early Muslim philosopher-scientists drew heavily from the body of knowledge they had appropriated from classical Greece, an intellectual legacy little known at that time to Western Europeans. Fundamentally, Muslims were guided by the teachings of the Islamic Revelation. Belief in the unity of all phenomena as set forth in the Qur'an, together with the classification of the sciences as, first of all, philosophical disciplines, promoted cosmological studies that, as a whole, reflected a broad range of approach. At one end was metaphysical and mystical speculation that extended beyond matters that could be revealed through direct observation or purely rational examination. At the other end was direct astronomical observation and mathematical analysis of the phenomena observed.

In exploring mystical Islamic cosmology, one must become familiar with a repertory of events and states of existence that are often presented in abstract terms such as pure being, essence, and absolute and infinite reality—all with special, esoteric meanings and dimensions beyond those that most Westerners today associate with space, time, and matter. Examining the concepts behind these terms is beyond the scope of this book, but they merit attention and study (see the list of Works Consulted at the back of this book). Virtually all of history's major cultures have contained such expression, and the Muslims inherited some of what they used from the ancient world. What seems common to much of the cosmology of pre-modern eras is a central philosophical concern with defining the place of human beings, together perceived as a microcosm, within the all-embracing universe, or macrocosm. Moreover, the cosmology of the past has at its core usually involved spiritual cause and ultimate purpose. In the twentieth century these elements have generally been absent from scientific effort as a whole. In medieval or "classical" Islam, however, as in other faith-dominated societies before and since, spiritual and metaphysical concepts provided stimuli and points of departure for speculation that was not inevitably at odds with practical experimentation and thus not irrelevant to scientific effort.

The guiding force of Islam's Revelation, with its transcendent view of all creation, did not prevent Muslim cosmologists from developing a number of different systems to explain the nature and workings of the universe. In particular, they focused increasingly on what ultimately became known as celestial mechanics. Here they were greatly influenced by *Al-Majisti* (the *Almagest*, or "Great Compilation"), the hugely influential astronomical text by the second-century Hellenistic Egyptian, Ptolemy. This work was based in part on Aristotle's earth-centered cosmos, and its mathematical concept of planetary movement in terms of spheres moving within spheres dominated the outlook of Muslim astronomers for centuries.

Less concrete and more mystical cosmic concepts also were developed. One of the most important of these was presented by Ibn al-ʿArabi, a twelfth-century Muslim Sufi mystic, teacher, and poet. In his view all phenomena are nothing but manifestations of Being, which is one with God. All emanate from the One, and there is no real difference between God and the universe. Ibn al-ʿArabi's beliefs also reflect a form of pantheism: he found deep spiritual value in Judaism and Christianity as well as in Islam.

Eventually, more specifically astronomical concepts of the cosmos were developed. It became a central effort among Muslim astronomers to reconcile Ptolemy's model of the cosmos with mathematical equivalents that described a cosmos more in accord with what they were coming to perceive as the actual positions and movements of the heavenly bodies. This reform of planetary theory was among the most important of the Muslim scientific achievements in astronomy. At hand and ready to help in this endeavor, in particular, was another vital part of the classical and Eastern legacy that the Muslims had appropriated and were soon to enhance significantly: mathematics, the fundamental tool and language of scientific investigation.

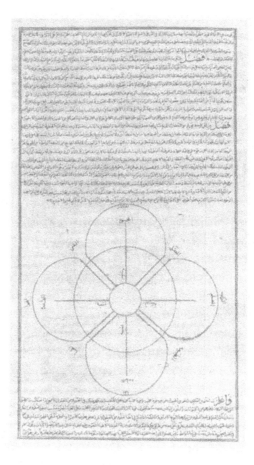

Figure 4.1
Diagram of Mystical Cosmos, MS Illustration from *Al-Futuhat al-Makkiya* (The Meccan Conquests), by Ibn al-ʿArabi, Sixteenth Century

The unifying force of the Islamic Revelation, with its transcendent view of all creation, did not prevent Muslim cosmologists from developing differing systems to explain the nature and workings of the universe. Here is a metaphysical concept of Ibn al-ʿArabi, one of Islam's greatest Sufi mystics. In his view, all phenomena are nothing but manifestations of Being, which is one with God; all emanate from the One. There is no real difference between God and the universe. A concept such as this reflects the conviction that one must go beyond the material or tangible appearance of material phenomena to perceive, through emotion and intuition, the true nature of ultimate, universal reality.

Figure 4.2
Man and the Macrocosm, MS Illustration from *Zubdat al-Tavariq*
(Treasure of History), by Loqman, Turkey, Sixteenth Century

Less esoteric than the Ibn al-'Arabi diagram (Fig. 4.1), this miniature painting includes a number of elements that have been associated with cultures beyond Muslim lands. Man is shown in the heavens, each ring of which corresponds to one of the prophets revered in Islam (including Abraham, Moses, and Jesus). Around the heavens circle the zodiacal signs and the lunar mansions. At each corner is a representation of the angelic realm, gateway to the Divine Presence. The concept of concentric paths or spheres long remained a staple of graphic presentations of the cosmos; a somewhat similar rendering of the universe appears in Renaissance paintings produced in fifteenth-century Siena.

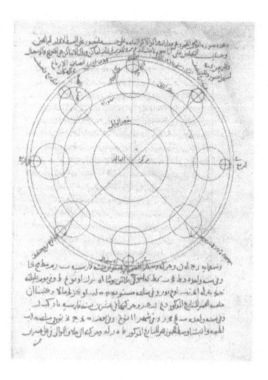

Figure 4.3
Diagram Relating to Ptolemy's Theory of Planetary Motion,
MS Illustration from *Nihayat al-sul* (A Final Inquiry Concerning the
Rectification of Planetary Theory), by Ibn al-Shatir, Fourteenth Century

Developed long before the Muslims' esoteric or metaphysical schemes, relatively rational concepts of the cosmos exercised profound influence on Muslim philosophers and astronomers. Many of these were derived from the celebrated text *Al-Majisti*, known in Europe as *Almagest*, which was written in the second century AD by the Egyptian astronomer Ptolemy and based in part on Aristotle's earth-centered cosmos. Conceived as actual spheres moving within spheres, this motion of planetary bodies was also defined in mathematical language by Ibn al-Shatir, a noted Muslim astronomer who, along with many of his contemporaries, devoted considerable time and effort to converting Ptolemy's cosmic model into a workable, equivalent concept more in accord with what could actually be seen in the heavens. There are interesting parallels between the work of Ibn al-Shatir and the sixteenth-century Polish astronomer Copernicus (see Fig. 6.31). Long indistinct, the boundary between cosmology and astronomy was gaining definition, as the profound appeal of the physical world and its environs was growing.

Mathematics

NATIVE TONGUE OF SCIENCE

The historic Islamic achievement in clarifying the intellectual legacy of the past, putting it in workable order, and then enriching it with significant innovation is notably evident in Muslim mathematics. To begin with, what mathematics had Muslims inherited and from whom? A great deal, it is clear, and from many sources, much of it having a sophisticated character that had been developing for four thousand years. The ancient Egyptians knew about decimal numbers, solved complicated problems through equations involving squared numbers, calculated circular and spherical areas quite accurately, and, in general, could apply the kind of mathematical skills required for such gigantic tasks as planning and building the pyramids at Giza. Measurement of geometric forms, both plane and solid, was known to the ancient Egyptians and Mesopotamians. The Sumerians, as early as the fourth millennium BC, appear to have employed an intricate system of accounting. The Babylonians were familiar with place value numerals, that is, with numbers whose value changed automatically and consistently according to their position within a figure—first, second, or whatever. Place value came to be used in numbering systems based on a number such as ten (the decimal system) or sixty (the sexagesimal system). The place value concept, while so fundamental as to seem now self-evident, was one of the most important achievements in the history of world science.

A thousand years before the Christian era, Chinese arithmetic could solve complex problems of surveying and measuring geometric areas. In India, as far back as the fourth century BC, calculation made use of an algebraic-like method inherited from Babylonia, and one system of numbers that foreshadowed modern Hindu and Arabic figures. Place value decimals became known in India around the first century AD, and numerals began to include a form of zero, which had evolved from a Babylonian symbol representing a blank space.

Mathematics thus became an international instrument of calculation long before the advent of Islam. It took the Greeks of the last few centuries before the Christian era to turn this instrument into a profoundly disciplined and resourceful language—a set of laws and terms that could be used to measure and reveal, with a precision and subtlety never before possible, the inherent order of everything in the natural, physical world.

From the time of Pythagoras, numbers and their relationships mesmerized the Greeks and, along with geometric shapes, allowed them to perceive a whole universe and, in a sense, to comprehend its structure and function. Not for nothing were mathematical disciplines built into Greek philosophy: mathematics, indeed, amounted to philosophical exercise, using established procedures of argument, demonstration, and proof to arrive at answers that would work universally. In the third century BC, Euclid, the greatest Greek mathematician of Alexandrian times, produced his *Elements*, which assembled in thirteen books all the geometry acquired up to that time. This work also covered number theory, irrational numbers, and other related subjects, treated in terms of definitions and axioms. Euclid's achievement was an important part of the heritage received by the Muslims, and it remained virtually unchallenged until the nineteenth century.

Greek geometry and Hindu arithmetic and algebra reached Muslim lands at an early date, the Greek discipline being part of the treasure of scientific manuscripts that were translated in centers such as Gondeshapur and Baghdad, the Hindu probably arriving through commercial links with India. Two fundamentally different ways of mathematical investigation thus came together during the early centuries of Islamic culture: the

Greeks' tendency toward visualizing concepts geometrically, and the Babylonian emphasis on sexagesimal (or base sixty) computation with place value numerals, along with the use of place value decimals appropriated from India.

Starting out at intellectual centers like the houses of wisdom set up by the Abbasid caliphs at Baghdad and the Fatimids at Cairo, the first Islamic mathematician-philosophers plunged into their intellectual booty avidly. They soon were familiar enough with it to begin criticizing those ideas, formulations, and details that they found inaccurate, inconsistent, or otherwise erroneous. They undertook new translations and revisions of existing ones, making corrections along the way and arriving at some new conclusions. Here must have been one of the great housecleanings of cultural history! In due course, Muslim mathematicians transformed the nature of numbers, streamlined some mathematical disciplines, and virtually developed a new branch of mathematics. Before looking more closely at the most important of these achievements, it is worth noting that in the same historical period Western Europe's mathematicians were spending their time in fussing with the calendar, updating instructions on using the abacus, and working still with Roman numerals.

Numbers

Figures using place value numerals inherited from Babylonia, together with the type of place value decimals that turned up in India early in the Christian era, became known to Islamic civilization in its early centuries and were recognized as incomparably efficient for all kinds of computation. Muslims inherited three systems of calculation: the first employed a finger-reckoning method of expressing numerical operations in terms of specific positions of the fingers, the results being written out in words; the second, from Babylonia, represented numbers by letters of the alphabet; the third made it possible to express any number by means of nine figures and a symbol, zero. The Arabs translated this symbol as *sifr*, or cipher, which indicated an empty place, a concept known to both Babylonians and Hindus. The numerals in this third system were Indian in origin, but no one has yet been able to determine exactly how they were transmitted to the Muslim lands. They came to be written somewhat differently in the eastern and western regions of Islam; it was the western type that was introduced to Europe and became known as "Arabic" numerals, the ones familiar to much of the world today. The very thought of using Roman numerals to try even the simplest calculation can only arouse our gratitude for the priceless and fundamentally simple numerical legacy passed on to us through such a complicated ancestry.

The Muslims were not only interested in using mathematics to solve problems of daily life; among philosophers and others given to speculation was a fascination for the Pythagorean concepts that were part of their classical intellectual inheritance. Interest in number theory, consideration of numbers as real things, exploration of magic squares and relationships between numbers and letters—all these activities gave Islamic mathematics an arcane and mystical edge, which extended into such areas as alchemy and magic.

Mathematical Subjects

The everyday problems to which the Muslims applied their enhanced mathematics naturally increased in scope and complexity as Islam extended its domain. The "arithmetic of daily life" was essential in everything from the calculation of taxes to the division of personal estates in accordance with laws set forth in the Qur'an. Fundamental mathematical principles and definitions, Greek in origin, were clarified by Muslims in such a way as to increase public understanding of numerical relationships and promote efficiency in all kinds of computation. At first sexagesimal notation predominated, but eventually it was replaced by decimal numeration, which made complex calculation much easier. Meanwhile, in the course of developing their arithmetical operations, the Muslims added irrational and natural numbers and common fractions to the Greek repertory they had inherited.

Muslim geometry descended directly from principles set forth by Euclid, and it showed Indian influence as well. In the ninth century, at Baghdad, the Banu Musa (three gifted brothers, sons of Musa), among other distinguished mathematicians, investigated problems in constructing interrelated geometrical figures. The character of the mathematical point, line, and space was given intense study, in a manner both mathematical and philosophical. Practical application was never long overlooked: Muslims utilized increasingly sophisticated geometry in surveying, in designing wheels of all kinds, including waterwheels and other systems for drawing water, in improving farming equipment, and, inevitably, in devising engines and devices of war, such as catapults and crossbows.

Great strides were made in theoretical work. The eminent eleventh-century mathematician Ibn al-Haytham delved into isometrics, the representation of figures with all edges drawn with true relative length, that is, without perspective distortion of dimension. He and many other noted mathematicians, such as Thabit ibn Qurra and Nasir al-Din al-Tusi, as well as Omar Khayyam (celebrated today much more as the poet who wrote the famous *Rubayyat* than as a scientist), devoted considerable time and effort to proving Euclid's Fifth Postulate: If a straight line cuts a pair of straight lines and produces two angles on the same side measuring less than ninety degrees each, it indicates that the pair of straight lines will ultimately converge. The complexity of the Fifth Postulate stimulated other Muslim mathematicians to arrive at alternate proofs—including proofs of non-Euclidean theorems. It was only much later, in the nineteenth century, that the medieval Muslims were understood to have defined what came to be called Euclidean geometry, and that, without realizing it, they had pointed the way toward the discovery of independent, non-Euclidean disciplines.

In general, Islam's mathematicians spent a great deal of time micro-examining: extending, putting in order, refining, and polishing their inheritance from Babylon, India, and ancient Greece. In algebra, however, they did more than clean house and make improvements. Here one man in particular stands out: al-Khwarizmi, a Persian born in the eighth century. Not only was he instrumental in converting Babylonian and Hindu numerals into a simple and workable system that almost anyone could use, he originated both the terms "algebra" and "algorithm," along with the concepts behind them. Algebra, or *al-jabr*, denotes a transposition—a restoration of balance (*muqabala*) or equilibrium—through adding or subtracting the same quantity on both sides of an equation; associated with this activity is reducing or simplifying and combining equivalent terms. Algorithm, the term itself derived in Medieval Latin from al-Khwarizmi's name, has come to denote any systematic computation or system of step-by-step instructions for solving a problem or pursuing some goal.

Al-Khwarizmi's algebra includes a clear demonstration of positional mathematics and examples of equations, as well as principles of the square root and other fundamental operations. In recognizing the capabilities of the equation in describing complex relationships through establishing balances, and in defining unknown factors, using a symbol such as x, al-Khwarizmi opened the door wide to the advanced mathematical procedures that became possible in later centuries. In sum, algebra's debt to Islam's mathematicians is very great—the discipline is in many ways their creation. Perhaps the most important difference between medieval Muslim algebra and our own is the way the former was written out, in words; the symbolic language we use was yet to be developed.

Trigonometry is also substantially a Muslim creation, especially in connection with studying plane and spherical triangles. It was given great impetus as a discipline by the requirements of astronomers concerned with mapping points in the sky on the celestial sphere. Trigonometry's functions, involving ratios such as sine and cosine, tangent and cotangent, were greatly developed and refined in the Islamic lands; Muslims replaced the ancient Greek system of chords (straight lines joining points on a curve), making it much easier to solve complex problems connected with spherical triangles.

Among the pantheon of illustrious Muslim mathematicians are many who, like the European Renaissance men of later centuries, also distinguished themselves in other sciences. Al-Khwarizmi was also an astronomer and geographer; Ibn al-Haytham was an astronomer and also specialized in optics; Abu Rayhan al-Biruni was a philosopher, historian, astronomer, pharmacologist, botanist, geologist, and mathematician, who, among other things, translated Euclid's works into Sanskrit and calculated the earth's circumference and radius with an accuracy not far from that of today's measurements. Historical periods characterized by especially significant cultural advance seem to produce the largest number of individuals spectacularly gifted in more than one science or art.

Music

According to the tales of the "Thousand Nights and One Night," <u>music is to some people meat, to others medicine.</u> Early Islam considered it to be an expression of ethics and of the harmony of the spheres. Such a view came naturally to a culture that had appropriated so much from the Greeks. For the great mathematician Pythagoras, born in the sixth century BC, music occupied a central place in philosophy. He also perceived precise mathematical relationships between musical notes, both in the way they looked and in the way they sounded. He found corresponding ratios between the distances between heavenly bodies and intervals of musical scale. He even explored acoustics. In addition, the basic Greek scientific curriculum grouped music with arithmetic, geometry, and astronomy. Thus, Muslims, adopting and adapting so much of their Greek legacy, quite naturally defined music as a specific branch of mathematics.

The actual music of ancient Greece had ceased to be heard long before the birth of Islam, but its theory was kept intact and transmitted through Byzantine and Persian channels. An extensive literature concerning Arab music began to be developed in the eighth century by musical scholars and specialists such as al-Mawsili and his son, both serving as official musicians at the brilliant Baghdad court of al-Ma'mun. Later writers on music such as al-Masudi and Abu 'l-Faradj al-Isfahani and giants of philosophy and medicine, such as al-Razi, al-Kindi, Ibn Sina, and especially al-Farabi, developed theories of music (even of music as therapy) and investigated the physical nature of sound itself. Most probed the nature and rules of musical forms. Al-Kindi examined what he perceived as the cosmological connections between the strings of the ʿud, a sort of lute, and such natural phenomena as the four elements, the seasons, and objects in the heavens. The tenth-century Ikhwan al-Safaʿ (Brethren of Sincerity), a remarkably liberal secret society of intellectuals and encyclopedic writers, also dealt with music in terms of cosmology, as well as numerology. Al-Farabi's *Kitab al-Musiqa* (Book of Music), considered by many the greatest work of its kind written up to that time, examined musical intervals and their combination. Both al-Farabi and Ibn Sina paid extensive attention to matters of rhythm.

Certain modes, types of melodies, and forms of ornamentation are fundamental to virtually all classical Islamic music. Melody maintains a principal position; there is little if any polyphony (counterpoint of independent melody), and ornamentation is marked with considerable nuance, complexity, and subtlety. Surface styles vary widely, reflecting the many races and peoples that Islam's civilization has embraced for more than fourteen hundred years. Yet even before one becomes familiar with this music, one's ear usually transmits to the mind what seems like a kind of mathematics in sound. Westernized hints of this may occasionally be heard in some of today's New Age music, with its sometimes mesmerizing repetitions carried forward at great length with slowly successive, minor variations. In any event, the influence of classic Islamic composers has long been clearly apparent in the songs of the troubadours of late medieval and early Renaissance Europe and in much of the music of Spain during the last thousand years.

MATHEMATICS

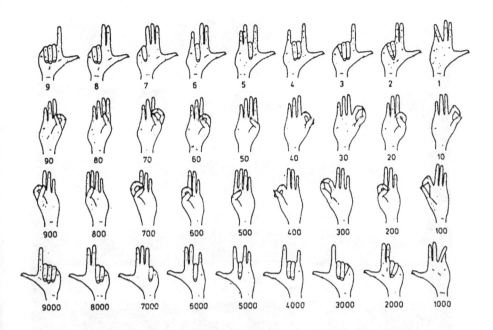

Figure 5.1
Diagram Showing a Demonstration of Finger Reckoning

Among the systems of calculation inherited by Muslims was a method of reckoning with the fingers, used generally when a scribe or clerk was calculating with numbers written out as words and needed to keep in mind the intermediary amounts he arrived at mentally. This task involved using both hands, the right manipulating the fingers to indicate units and tens, the left indicating hundreds and thousands. Any number between 1 and 9,999 could be defined in this way. Roman numerals appear to resemble somewhat the visual display produced by the fingers—I, II, V, and so on. The Hindu numerals and reckoning methods that were imported and adapted by Muslims during Islam's early centuries largely replaced earlier systems of reckoning. However, in a traditional Muslim marketplace today, the visitor may still come upon finger reckoning, sometimes carried on with the speed of
a skilled abacus user.

SCIENCE IN MEDIEVAL ISLAM

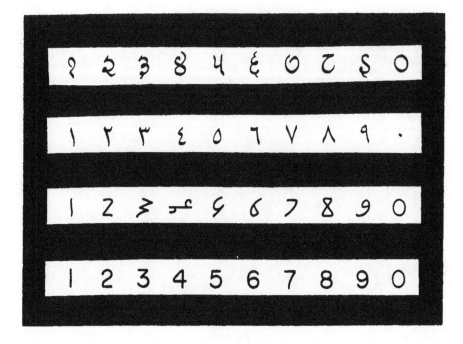

Figure 5.2
Diagram of the Development of Arabic Numerals, showing Indian
Sanskrit, Eastern Arabic, Early Western Arabic, and Modern Figures

In addition to the finger-reckoning method of expressing numbers, Muslims also inherited and adapted two other systems of calculation: first, an ancient Babylonian system that represented numbers by letters of the alphabet; and second, a superior system that made it possible to express any number by means of nine figures and a symbol, zero, which indicated an empty place. The figures were Indian in origin, and how they were transmitted to the Muslims is not fully known. They were written somewhat differently in the eastern and western Muslim regions. It was the western type that was introduced to Europe around the tenth century and became known as Arabic numerals. Shown above are (from the top): tenth-century Sanskrit, modern Arabic (eastern), early Arabic (western), and modern Arabic (western).

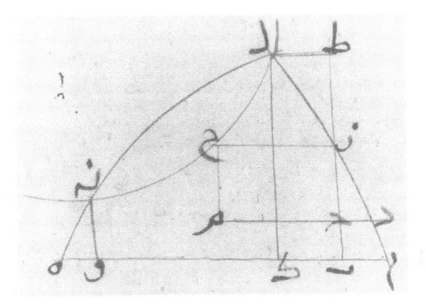

Figure 5.3
Demonstration of a Trinomial Equation by the Mathematician
Who Was Also a Poet, MS Illustration from a Text on the Algebra
of Omar Khayyam, Copy, India, Fourteenth Century

Omar Khayyam has long been celebrated chiefly as the author of the famous *Rubayyat*, which has enchanted readers around the world for more than half a millennium. Medieval Islam regarded this twelfth-century Persian polymath chiefly as a mathematician. He not only made use of Euclid's geometry to formulate and solve important new mathematical problems but also made significant advances in algebraic calculation, as well as important improvements in the calendar. Medieval Muslim mathematical innovation reflects a historic Islamic tendency to clarify and systematize past and present knowledge.

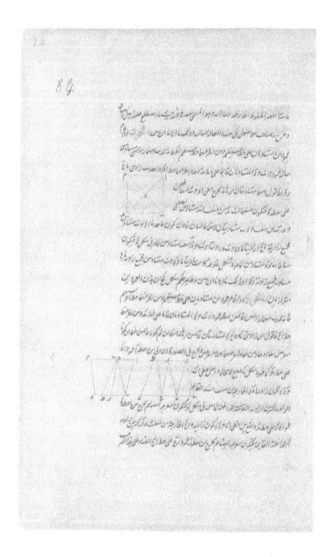

Figure 5.4
Proof of Euclid Postulate, MS Illustration from *Al-Risala al-Shafiya*
(The Satisfying Thesis), by Nasir al-Din al-Tusi, Thirteenth Century

Great attention was paid by more than one of Islam's best theorists to the postulates arrived at by Euclid in the third century BC. Shown here is the mathematician-astronomer al-Tusi's thirteenth-century demonstration of a postulate equivalent to Euclid's Fifth Postulate.

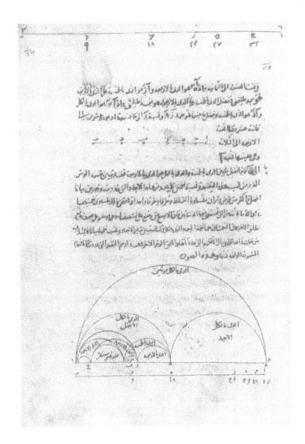

Figure 5.5
Mathematics and Muslim Music: Diagram Showing the Division of a
Musical Chord, MS Illustration from the Preface to *Kitab al-Musiqa*
(Book of Music), by Abu Nasr al-Farabi, Copy, Fifteenth Century

Pythagoras and later Greek scholars carried out lengthy investigations of the mathematical relationships in music, noting that musical notes differ in pitch according to specific ratios, which can be measured in the relative tensions of vibrating strings, as well as in different lengths of reed pipes. They were convinced that music expresses the kind of intrinsic order and harmony that is found in mathematics. The Pythagoreans even proposed that intervals of the musical scale reflect the same numerical ratios as the distances separating heavenly bodies and the central cosmic fire. All these concepts influenced Muslim musicologists, especially the philosophers al-Kindi, Ibn Sina, and al-Farabi.

Figures 5.6a, 5.6b, and 5.6c
Mathematical Configuration in Islamic Art

Four decorative elements—calligraphy, floral and vegetal forms, human and animal figures, and geometry—dominate virtually all the visual arts of the Islamic world. Not the least of these four is geometric pattern, which is found in all kinds of two- and three-dimensional items, from the smallest Persian and Indian miniature paintings and manuscript illustrations to the vast stone and ceramic-tiled domes and walls of great Persian mosques. The three illustrations of mathematically configured art that follow here may serve to suggest the extraordinary range and depth of the aesthetic, logical, and spiritual perception that Muslim designers and artists of varied media applied to their works. Such geometrical conceptions, whether developed simply or with the greatest sophistication, were frequently displayed in combination with other elements. No other culture has employed precisely geometric abstract shapes to such an extent and with such mesmerizing effect, inevitably leading the viewer's eye and mind beyond the finite realities of actual space into complex realms of seeing and reflection that are infrequently evoked in much Western art. There is an almost physical parallel between the character of these controlled visual pyrotechnics and the intellectual, cosmic, and harmonious elegance inherent in mathematics as a science—a field of scientific endeavor that medieval Muslim scholars associated with music and astronomy and placed within the category of the philosophical sciences.

MATHEMATICS

Figure 5.6a
Photo of Geometrical Pattern in Ceramic Tile, Courtyard Wall,
Alhambra Palace, Thirteenth–Fourteenth Centuries, Granada, Spain

Figure 5.6b
Photo of Stucco Stalactite Cupola, Chamber of the Two Sisters,
Alhambra Palace, Thirteenth–Fourteenth Centuries, Granada, Spain

Figure 5.6c
Photo of Ceramic Plate with Geometric Design, Morocco,
Nineteenth Century

6

Astronomy

Ancient Legacies

Medieval Muslims considered astronomy to be one of the mathematical sciences. Their efforts in this discipline consisted largely of studying the apparent movements of heavenly bodies and recording, in mathematical terms, what they found. Phenomena such as starlight and objects such as meteors and comets were assigned to the realms of physics and metaphysics, as was the fundamental nature of the heavenly spheres. Such classification was hardly Islamic in origin: what Muslims inherited in the way of astronomical concepts, terms, and practice can be traced back to ancient Greek and Ptolemaic astronomy; some knowledge also came from Indian and Sasanian sources. Included in this rich legacy was knowledge passed on over thousands of years from ancient Babylonia and Egypt, where observations of the sky were carried out in some detail and calculations of the calendar year were based on what could be seen in the skies—calculations that were often not far from those of modern times.

The Greeks greatly extended what they had received from the Babylonians and Egyptians, especially in the field of theoretical astronomy, which is concerned with the development of mathematical models of planetary position and motion. Virtually all the intellectual giants of Hellenic philosophy and science took part in pushing back the frontiers of the heavens as perceived from the earth. In the early centuries of classical Greece the boundary between cosmology and astronomy was unclear. The mathematician Pythagoras conceived a universe of heavenly motions divided according to degree of noble perfection: lowest was the earth and the sphere beneath the moon; above spread a cosmos framed in a sphere of fixed stars; farthest out was the home of the gods, Olympus. As has already been indicated, spheres were set neatly within spheres, everything moving in unchanging circles—a fundamental concept that was to imprison astronomers almost until modern times.

One Greek savant after another took his own particular look at the heavenly spheres and what went on within and around them. For some time the most widely accepted picture placed earth at the center of all the spherical motions; more than two dozen spheres came to be accepted. Eventually, the Pythagoreans' idea of a great central fire replaced earth as the center of the universe, with the sun moving around the fire along with the earth and the other planets. Aristarchus of Samos, who was active in the third century BC, was probably the first to propose a heliocentric, or sun-centered, system. Not until nearly eighteen hundred years later would Nicolaus Copernicus, Johannes Kepler, and Galileo Galilei build on this revolutionary idea and arrive at the fundamental structure and motions of the solar system that are generally accepted today.

Plato and Aristotle refined the mechanisms of movement within the Greek cosmos of spheres to the point where forces of downward and upward (or inner and outer) movement worked in a tense balance and where the four elements were assigned places in a hierarchy of nobility (with fire the most noble and earth the least). Heavenly bodies, far nobler than terrestrial ones, were still considered the more perfect the farther out lay their paths. Greek philosophers generally agreed that the universe operated in an orderly way which was teleological: it reflected a fundamental purpose. In assigning every part of the universe its proper place, they defined two regions, one, with terrestrial matter, extending below the moon, the other, with celestial matter, stretching beyond; each region was ruled by different principles. Gone were the fanciful angels and other mythic guardians of more ancient cosmos: here was an austere division of inanimate geometric forms moving in ceaselessly unchanging, rigidly circular paths. Here was theory at its most drily theoretical. In time, it began to look less and less scientific and to seem less and less intellectually acceptable.

The concept of the planetary system bequeathed by the Greeks to the civilizations of the early Middle Ages continued to place the earth at the center, with the heavenly bodies moving around it in the following order: nearest was the moon, then Mercury, Venus, the sun, Mars, Jupiter, and Saturn, with the fixed stars farthest out. However, the Hellenistic astronomer Aristarchus looked at things differently: he believed that the earth made a full rotation on its axis every day and orbited around the sun once every year; the sun and fixed stars did not move. Aristarchus' contemporaries would have none of this, as it challenged their belief in the different natures of the earth and heavens.

The earliest Islamic astronomers were influenced by Indian and Sasanian texts. The royal court of the Sasanid empire at Ctesiphon sponsored broad scientific investigations between the third and seventh centuries, particularly in astronomy and medicine; a second center was developed at Gondeshapur, in Persia. Later, one of the great early caliphs of the Abbasid dynasty, al-Mansur, assembled a group of Persian, Indian, and other scientists at Baghdad, and by the eighth century the internationalization of Islamic science was well on its way. The son of the eminent ninth-century translator Hunayn ibn Ishaq produced his own version of Ptolemy's *Al-Majisti*. Virtually all of Islam's astronomers accepted its theories as the definitive mathematical model of the heavens, with the sun and planets revolving around a fixed earth in combinations of circular orbits.

An early Alexandrian astronomer, Apollonius, working during the third century BC, had suggested that variations in the distance of a planet from the earth could be explained by assuming that a planet moved in a circle, called the epicycle, whose center traveled around another circle, the deferent, centered on the earth. Fig. 6.5a indicates the proposed patterns of these circular motions. Another concept, suggesting that planets moved in eccentric circles around the earth, with the orbital centers situated some distance from earth's center, was used by the Alexandrian astronomer Hipparchus, who explained the apparent motions of the sun in terms of a fixed circular orbit eccentric to the earth, using epicycles to describe the orbits of the planets. Ptolemy adopted both epicycles and eccentric orbits in formulating his model of the heavens and their bodies; by this time it took as many as forty-one circles to accommodate all the activity taking place in the skies.

Most medieval Muslim astronomers never abandoned the fundamental views of the heavens set forth by Aristotle and Ptolemy. The vision of firmly nesting spheres, one moving around another, continued to frame most thinking about what went on beyond Earth, whatever the elaboration and complication introduced by epicycles and eccentric orbits. By and large, the astronomers were cautious. They followed Ptolemy in avoiding any definition of the overall nature of the heavens: that was a matter for philosophers and metaphysicians. Mathematics provided the proper theoretical approach to astronomical celestial matters. The Aristotelian concept of solid spheres that had been introduced to Muslims through the works of Ibn al-Haytham remained for centuries the fundamental model. Under this formulation, spheres and celestial bodies were composed of one substance, a fifth element, essentially different from fire, air, earth, and water. The stars were kept in place by the spheres, which, rotating, drew the stars along with them. Such invisible mechanics seemed ordained; they also provoked continual scholarly observation, calculation, argument, and revision.

Ptolemy's *Al-Majisti*, particularly its scheme of heavenly motions, called kinematics, was destined to dominate astronomical thought throughout the Islamic world and Western Europe until the sixteenth century. By the eleventh century, astronomy was flourishing in the Western Caliphate in Spain, where, at the great cultural center of Córdoba, Ptolemaic astronomy began to be questioned, in both philosophical and physical terms. This disputation was also in progress among astronomers in the eastern Islamic regions. Despite continued allegiance to Ptolemy's main propositions, Muslim astronomers occupied themselves for several centuries in trying to make his model more compatible with what their eyes could tell them. In so doing, they fiddled, calculated, recalculated, made new observations, reconciled them with the old ones (or adjusted the old ones), all in the cause of gaining accuracy and precision. Meanwhile, in thoroughly practical ways, and from the very start, they applied what they could see, calculate, and record to help them in practicing their faith.

Science Serving the Faith

Long before Muslim astronomers developed their most sophisticated observational and theoretical methods, they became skilled in applying astronomical knowledge to meeting the fundamental requirements for worship.

ASTRONOMY

Islamic religious practice has always necessitated precise determination of time and place, whether in connection with prayer or with determining the beginnings of months and holidays in the Muslim lunar calendar.

Muslim prayers must be oriented in space as well as scheduled in time. All Islam prays toward Mecca, the ancient city which contains the most holy of all Muslim shrines, the Ka'ba. In almost every mosque throughout the world the *mihrab*, or prayer niche, is placed so as to turn all worshipers in the direction, *qibla*, of the Ka'ba. This orientation is often apparent in the way each mosque structure is sited.

A branch of astronomy that medieval Muslims termed the *'ilm al-miqat*, the science of timekeeping, also known as the "science of the fixed moments," was applied through direct and instrumental observation as well as mathematical calculation in order to fix the five times of daily prayer: sunset, late evening, dawn, soon after midday, and late afternoon. The Islamic day begins at sunset; so does the Islamic month, on the day and at the hour the new moon is first seen. Insofar as fixing times for prayers and for the calendar involved watching the heavens, the basic procedures were not new when Islam was founded; they were known to the ancient Babylonians and Egyptians.

Moreover, the Arabs had studied the night skies for centuries in order to mark the time that passed during their long treks across the desert. They could recognize the location of specific groups of stars, as well as the stages and relative positions of the moon, as indicators of the seasons. Such navigational and calendrical guideposts were adapted and refined to accommodate the requirements of Islamic religious practice. A trio of astronomical and mathematical disciplines was pressed into this service: such application has virtually no parallel in the sciences of ancient Greece or medieval Europe. It was an unrivaled and increasingly sophisticated effort, as an enormous body of recorded observations and calculations testifies.

The investigation of the celestial sphere—that apparent dome, filled with heavenly bodies, that we perceive from earth—involves spherical astronomy and mathematical geography, both of which, along with increasingly complex mathematics, were employed by Muslims in determining the precise moments in time and degrees of geographical direction that are essential to worshipers. At first, observation of the lengths of shadows cast by the sun was used to regulate prayer times; later, tables were calculated, correlating shadow lengths and the height of the sun and indicating the lengths of the intervals between prayers. Instruments were developed to indicate the local direction of Mecca. Official timekeepers, *muwaqqit*, employed by the mosques, set (and had their muezzin call out, usually from the minarets) the times of prayers according to their own or someone else's records of observations and calculations. Sometimes the timekeepers were professional astronomers. Their records were prepared in accordance with the astronomically based science of timekeeping already noted and, in the form of tables or almanacs, were eventually produced in great quantity throughout the Islamic lands. They became increasingly accurate and comprehensive as the available instruments for observation and calculation were replaced by more sophisticated sundials, quadrants, astrolabes, compasses, and *qibla* indicators.

The administrative and communication needs of Islam's early expansion made a new, particularly Islamic, calendar necessary, so a reigning caliph in the seventh century established one that, unlike the Julian and Gregorian calendars, was based on lunar rather than solar cycles. This new calendar began on the first day of the year (AD 622) of the Hegira, the emigration of the Prophet Muhammad from Mecca to Medina. This date, estimated to have occurred in late September, marks the start of Year One in the Islamic calendar. The fact that the calendar was based on a strictly lunar year has made conversion between the Islamic and Gregorian calendars a somewhat complicated procedure, as is described in the Introduction to this book. All Muslim holidays and festivals, as well as Ramadan, Islam's annual month of fasting, are scheduled according to the lunar month. Thus, the first actual sighting of the crescent of the new moon is an important moment for all practicing Muslims.

Observational Astronomy

Muslims began organized and detailed observation of the skies soon after the early expansion of Islam. This effort was naturally accelerated by an increasing demand for precise tables needed in preparing calendars, prayer tables, and horoscopes. A number of observatories, usually founded or sponsored by caliphs and other rulers, were established at centers such as Rayy, Isfahan, and Shiraz in Persia, as well as in Egypt. Very often the establishment of an observatory was prompted first of all by royal interest in astrology.

Two particularly impressive observatories, with large professional staffs, were established, one in the thirteenth century at Maragha, in Persia, the other in the fifteenth century at Samarkand, in what is now Uzbekistan. These centers were built by Mongol and Turkic rulers, descendants of Genghis Khan and Tamerlane respectively, whose forces invaded Muslim territory from the east in the thirteenth and fourteenth centuries, conquered vast areas of Southwest Asia and Asia Minor, converted to Islam, and established new and powerful dynasties. At these observatories new readings mapped the skies in unprecedented detail, producing a comprehensive and remarkably accurate picture of the stars and their constellations, thus providing an invaluable framework for observations carried out by later generations of astronomers in both the East and West.

Somewhat as do our eminent universities today, medieval Islamic observatories and royal courts attracted celebrated scholars and teachers, who served as magnets, attracting other scholars as well as students from all the Muslim regions. The celebrated astronomer Thabit ibn Qurra was associated with the venerable observatory established at Baghdad in the ninth century by the great Abbasid caliph and patron of learning, al-Ma'mun. Tenth-century astronomer and mathematician al-Battani had his own private observatory at Rakka, Syria. The eminent eleventh-century Egyptian astronomer Ibn Yunus led observational studies at Cairo. While attached to the court observatory at Ghazna, in Afghanistan, al-Biruni, an astronomer, mathematician, natural historian, and pharmacologist, provided observational data that formed the basis for important astronomical tables, known as *zij*. In addition, the celebrated houses of wisdom at Baghdad and Cairo played their part in early Muslim astronomical progress, developing a variety of tables from data collected from the observatories. Al-Khwarizmi, the great mathematician-astronomer, took part in this work during the time of al-Ma'mun.

Muslim Astronomical Instruments

Refinement in determining the lengths of the seasons, increased detail of solar and planetary motion, more exact terrestrial location of the important cities and towns—all these and more were achieved through Muslim observational astronomy, thanks not only to the skill and diligence of the observers but also to the increasing quantity and quality of the Muslims' observational instruments.

Most spectacular of all medieval astronomical apparatuses were the very large outdoor observational structures—actually, instruments—built partly underground at such observatories as those at Maragha and Samarkand, later at Delhi and Jaipur in Mughal India, and in Turkey. The greater the dimensions of the various devices, the greater the precision that was achieved in determining the positions of celestial bodies at various times of night or day. Few traces of the earliest structures remain, but the ruins that are to be found are impressive. Many of the instruments at Delhi and Jaipur are intact. Today's visitor can climb about and examine the ways in which astronomers could read the various sundials and other calibrated instruments and not only determine the altitude and azimuth of the sun and other stars but also convert the night ascension and declination of celestial bodies into latitude and longitude, as well as perform other operations aimed at mapping the changing sky throughout the year. Ironically, the eighteenth-century structures built in India under the sponsorship of the Muslim, or Mughal, Emperor Jai Singh became obsolete soon after they were put into use. The telescope, first produced early in the seventeenth century in Europe, was quickly exploited successfully, most notably by Galileo.

It was in the development and improvement of much smaller astronomical instruments that the Muslims made the most important advances. A particular case is the astrolabe: here, without question, was the most important computational instrument of the Middle Ages and early Renaissance. Probably a Greek invention of about the second century BC, the astrolabe was significantly enhanced—one might say perfected—by Muslims. This marvelously compact, often very small, device functions like a sophisticated version of an engineer's slide rule, or even like an analog computer, in solving a great variety of astronomical and time-keeping problems. In addition to determining the five Muslim prayer times and the local direction of Mecca, the medieval astrolabe, with its replaceable plates, calibrated for use in different geographical locations, could be manipulated to provide many kinds of year-round celestial and timekeeping data, terrestrial measurements, and astrological information. Introduced to Europe in the late Middle Ages, it was the subject of many texts, including a celebrated essay by Geoffrey Chaucer. A fine astrolabe almost always displays extraordinary craftsmanship. Medieval Islamic astrolabe-makers took understandable pride in their brass and bronze creations, the finest of which almost always bear the craftsman's name.

Second only to the astrolabe in importance was the astrolabic quadrant, a very simplified version of the astrolabe. Uncomplicated in construction, the quadrant, shaped like a ninety-degree pie segment, could be used to solve all of the standard problems of spherical astronomy (problems related to mapping features of the celestial sphere) for a particular latitude. Developed by Muslims in Egypt in the eleventh or twelfth century, it had, by the sixteenth century, replaced the astrolabe everywhere in the Muslim world except Persia and India.

Another non-observational instrument, the celestial globe was sometimes prized for its beauty as much as for its scientific utility. Basically a teaching device, the globe was used mainly in demonstrating the apparent daily rotation of the celestial sphere (representing the universe) over a horizon, represented by a ring within which the globe could be adjusted to reflect any terrestrial latitude. Celestial globes came also to be prized as ornamental gifts, lending to a caliph's office the distinctive touch that is often furnished today by the ship or aircraft model in an executive office.

Telling time by measuring shadows, the long-established sundial, dating from Greek and Roman times, was adapted to Islamic usage by incorporating curves that would define the "arrival" of shadows that indicated prayer times while the sun was up. Sundials may still be found set into the structure of very old mosques. Another category of instruments included devices primarily used for observation rather than computation. Chief among these was the armillary sphere, a physical representation of important features of the celestial sphere (see Fig. 6.15).

Celestial globes, astrolabes, quadrants, and sundials evolved in a variety of ways, and once the compass had found its way to Muslim lands it was adapted in several forms. In particular, *qibla* boxes, indicating the direction of Mecca, became widely used in Ottoman times after the thirteenth century.

Theoretical Astronomy

The Ptolemaic planetary system inherited by Islamic philosophers and astronomers accepted the principle of uniform circular motion while allowing the planets to move in epicycles, as was originally conceived by the Greeks. A concept of planetary motion involving epicycles had long intrigued and confounded the astronomers of ancient times. There had to be found a satisfactory way of explaining such commonly observed phenomena as the precession of the equinoxes, variations in the apparent sizes of the planets, and the retrogradation or apparent backing-up of a planet in its course around the heavens. Long considered as wandering stars, planets, moving slowly from west to east across the celestial sphere, have a way of occasionally appearing to stop and back up in relation to the fixed stars around them, then once again moving forward as before. Greek and Indian astronomers struggled with this phenomenon and made some modifications in their models of the heavens to accommodate it. Ptolemy refined the epicycle-deferent mechanism by adding the device of an equant, an eccentric or off-center point around which circles the large circle, or deferent, on which is centered the epicycle that indicated the planet's path. The equant could account for the apparent approach, recession, and backing up of the planet (see Figs. 6.5a and 6.5b). It represented the most sophisticated attempt made up to that point to square what the eye could observe with the way in which Ptolemy's theory stated that a planet must move.

Despite their general allegiance to Ptolemy's cosmos, Muslim astronomers eventually came to object in particular to the way his epicyclic motions violated the principle of uniformity of motion. This principle had been central to Greek and Indian physical concepts of all heavenly bodies in the universe and was firmly accepted by Ptolemy and most other later astronomers, Muslims included, until the sixteenth century and the findings of Kepler. This objection ultimately brought about a very important reform of planetary astronomy, effectively initiated in the thirteenth century by al-Tusi, the Persian mathematician, astronomer, and astrologist. His famous concept, known as the Tusi Couple, presented a hypothetical model of epicyclic motion that involved a combination of motions each of which was uniform with respect to its own center (Fig. 6.30). This model was applied to the motions of all heavenly bodies in the fourteenth century by the astronomer Ibn al-Shatir, who served as *muwaqqit* at the Great Mosque of Damascus. Thanks to the innovative refinements it contained, Ibn al-Shatir's formulation came closer to integrating observational and theoretical astronomy than had any other model of planetary motion up to that time. Slowly but surely the long-established canons of classical astronomy were being challenged by Muslim attempts to force Aristotelian and Ptolemaic theories into a practical, functioning system that described what really went on in the space surrounding the earth.

The way was slowly being cleared for a new kind of astronomy in which theory and observable fact could be reconciled without having to accommodate noticeable discrepancies. Some historians of science think it possible, even probable, that the Polish astronomer Copernicus, visiting the Vatican Library in Rome, saw Ibn al-Shatir's fourteenth-century manuscript illustrating his concept of planetary motion. In any event, a diagram in Copernicus' *Commentariolus* (ca. AD 1530) bears a remarkable resemblance to Ibn al-Shatir's schematic. Here may have been one of the most pivotal connections in the history of astronomy! Ibn al-Shatir's concept was conceived as an important element within what was believed to be an earth-centered planetary system. Copernicus' conceptualization of the same kind of motion fits perfectly within his own sun-centered planetary system, which has come to be accepted throughout the world as the true picture.

Whether or not there was a direct connection between the concepts, such Muslim innovations in astronomical theory amount to a major step in the historical development of astronomical science. They expanded the pursuit of knowledge in ways that helped to generate new methods of inquiry that were to evolve and flourish in the European Renaissance and Enlightenment. It is perhaps not surprising that al-Tusi, the thirteenth-century Persian astronomer-mathematician-astrologer who devised his celebrated "couple," was the first to treat trigonometry as a separate discipline, independent of spherical astronomy. It enabled astronomers to compute distances and directions of points on the celestial sphere more efficiently and precisely than ever before.

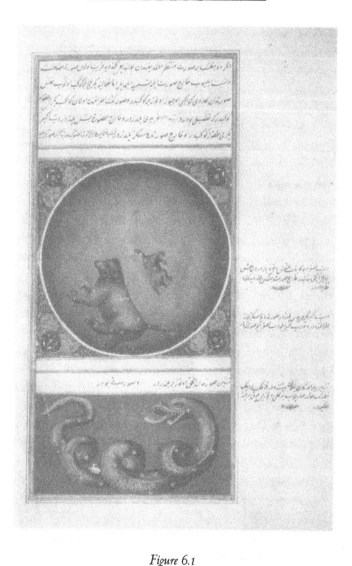

Figure 6.1
The Constellations Little Bear, Great Bear, and the Dragon, MS Illustration from *Iqd al-Djuman fi ta'rikh ahl al-Zaman* (General History), by Abu Muhammad Mahmud ibn Ahmad ibn Musa Badr al-Din al-'Ayni, Fifteenth Century

The celestial population was a favored subject for Muslim illustrators of every period and style. This Turkish painting decorated a copy of a general historical survey by a noted fifteenth-century scholar, al-'Ayni.

ASTRONOMY

Figures 6.2 and 6.3
Manuscript Illustrations from *Kitab Suwar al-Kawakib al-Thabita* (Treatise on the Constellations of the Fixed Stars), by 'Abd al-Rahman al-Sufi

The tenth-century Persian astronomer al-Sufi became celebrated for the precise observations and descriptions of the heavens collected in his most famous work, *Kitab Suwar al-Kawakib al-Thabita*. It included data on the forty-eight Ptolemaic constellations, tables showing the positions, magnitudes, and colors of their stars, and two views of each constellation, one as it appeared to an observer on earth, the other as it would appear on the celestial globe as seen from outer space. This unprecedented astronomical encyclopedia was based on observations recorded in earlier star catalogs, including those prepared by Ptolemy. In addition to providing scientifically important data, the al-Sufi manuscripts display, through the elegant paintings that accompany each celestial chart, the talents of some of Islam's greatest illustrative artists.

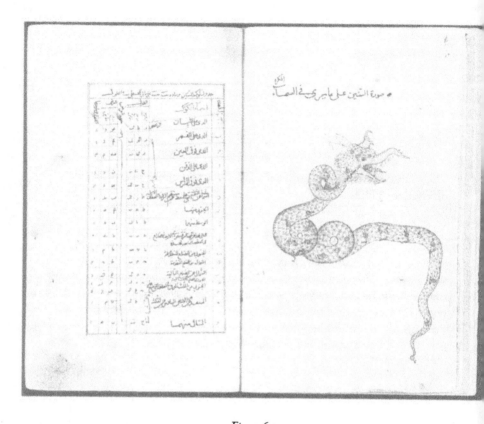

Figure 6.2
The Constellation Draco, Copy, Fourteenth Century

Figure 6.3
The Constellation Sagittarius (The Archer), Copy,
Persia, Seventeenth Century

Figure 6.4
Page (Detail) from a Ninth-century Greek Manuscript of
Ptolemy's *Al-Majisti*

Around 140 AD the Egyptian astronomer Ptolemy wrote a theoretical text that set forth mathematically the positions and movements of the planets in relation to the fixed stars. Its data were based on his own and earlier direct observations of the heavens. Called *Al-Majisti* by the Arabs, this Hellenistic work was the most systematic and comprehensive handbook of the heavens in use during ancient and medieval times, and it had almost as much influence on later generations of scientists and scholars as did Euclid's mathematical texts. Muslim astronomers eventually made many modifications and corrections to the text, but they rarely escaped its influence.

ASTRONOMY

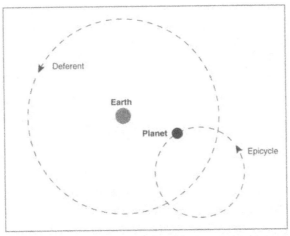

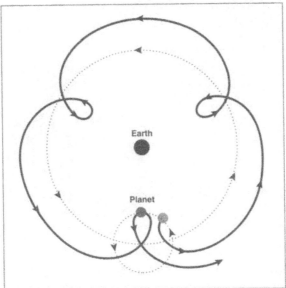

Figures 6.5a and 6.5b
Diagrams Illustrating Epicyclic Planetary Motion
(Epicycles and Deferents)

Apollonius, an astronomer active in the third century BC, suggested that variations in the distance between a planet and the earth could be explained in terms of epicycles and deferents. The diagrams above indicate these elements and circular motions.

Figure 6.6
The Astronomy Lesson: Teacher of Astronomy with Students,
MS Illustration from Persia, Fifteenth Century

The classroom pictured here was probably inside a royal palace or princely school. The teacher may have been either a celebrated visiting astronomer or perhaps the local royal astronomer-in-residence. The students were quite likely members of a small and elite body whose special abilities had earned them the privilege of studying from a master; almost certainly some of them had come thousands of miles for this privilege. During medieval Islam's most intellectually active centuries the special disciplines of advanced science were usually taught in an exclusive environment.

ASTRONOMY

Figure 6.7
Photograph of the Ka'ba, Mecca, Saudi Arabia

The Ka'ba at Mecca, Saudi Arabia, with the Sacred Mosque that encloses it, constitute Islam's holiest site. All Muslims are required to make a pilgrimage to it once in a lifetime if it is physically and financially possible. According to Muslim belief, the Ka'ba was built by the Prophet Abraham, and the site was a center of worship and pilgrimage for thousands of years before Islam began. After conquering Mecca in the seventh century AD, the Prophet Muhammad destroyed the ancient idols within the Ka'ba and established its central position in Muslim religious worship, redirecting the direction of prayer from Jerusalem to Mecca. Since that time, the world's Muslims have turned in prayer toward this place five times each day, and mosques in all lands have been oriented in its direction. The site is the most conspicuous orientation point marked or depicted on many Muslim compasses and related directional instruments.

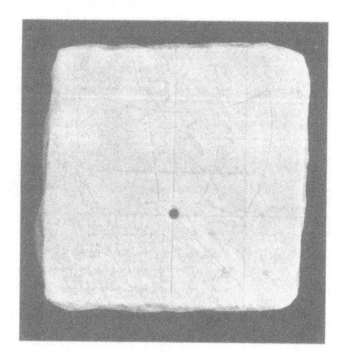

Figure 6.8
Photograph of an Arab Sundial, Horizontal, Stone,
Tunisia, Fourteenth Century

Dating back to antiquity, the sundial was investigated in detail by the Arabs, who made many improvements in the instrument's versatility and accuracy in the course of adapting it to serve the requirements of Islamic worship. Among the many scientists who devoted attention to sundial design was the mathematician al-Khwarizmi, who prepared tables for constructing sundials for use in different latitudes. The fourteenth-century example shown here displays the times of the daily prayers, which are indicated as the changing shadow cast by the gnomon moves across lines, known as traces, incised in the stone. In medieval times, the time of day in Islam was usually reckoned in seasonal hours, or twelfth divisions of the length of daylight. In this example, the five times of religious significance are shown by the five curves that intersect the three more or less horizontal "shadow" traces. These three traces correspond (from bottom to top) to the summer solstice, when the shadow is shortest, the equinox, and the winter solstice, farthest from the gnomon, when the solar meridian altitude is at its minimum and the corresponding shadow is longest. The *qibla*, the direction of Mecca, from Tunis is indicated by the diagonal line in the lower right corner.

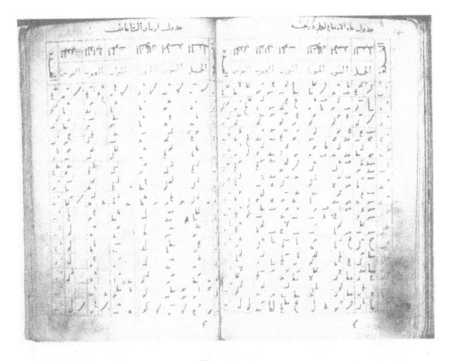

Figure 6.9a
Timetables and Orientation for Prayer (*Qibla*), MS Illustration from
Main Corpus of Tables Used for Timekeeping in Medieval Cairo,
Late Tenth Century

Astronomical determination of time and place in connection with religious practice has always been the most Islamic manifestation of Islamic science. The need to determine the local times of prayer and the *qibla*, the direction of Mecca, which all Muslims must face when they pray, and the need to calculate the beginnings and ends of the months and holidays in the Muslim lunar calendar, were met by widespread production of prayer and *qibla* tables such as those shown here, used in keeping time by the sun. Handbooks containing such tables are known as *zij*. The manuscript illustrated here contains some 200 pages of tables, with a total of about 35,000 entries, given in degrees and minutes of angular measure. All the tables in this work were computed for the latitude of the Cairo area and are attributed to the tenth-century Egyptian astronomer Ibn Yunus. Most of the tables were concerned with problems of spherical astronomy that involved reckoning time from the sun's altitude, as well as determining the sun's direction. The remaining tables were used to find the times of prayer, according to each degree of solar longitude (corresponding roughly to each day of the year).

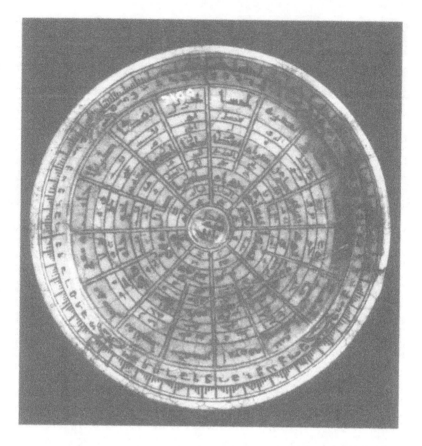

Figure 6.9b
Ceramic Plate ("Mecca Plate"), Made by Salim Thabit al-Dimashqi,
Syria, Thirteenth–Fourteenth Centuries

This unique plate was made, as one of the inscriptions indicates, for a king, "the victorious, the just al-Khagan, the son of the sultan, the king of the two ends of the earth and the two seas, the servant of the two sanctuaries, the honored, may God make his kingdom everlasting." It displays the kind of information usually found in medieval Muslim astronomical handbooks or on instruments such as astrolabes. Mecca is indicated at the center. The names and *qibla* directions for forty-eight localities are included, the directions being given in degrees and minutes from due south; the quadrant of the *qibla* (northeast, southwest, and the like) is noted as well.

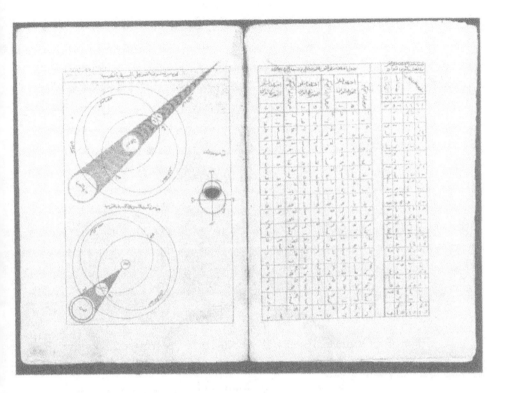

Figure 6.10
Pages from a *Zij*, an Astronomical Handbook with Text and Tables,
by Ibn al-Shatir, MS Illustration, Turkey, Sixteenth Century

Observations of eclipses and the positions of planets and stars were recorded in astronomical handbooks such as the one illustrated above. These records were often collected at medieval Islam's observatories with the use of the sophisticated instruments that Muslims developed in the Middle East, Spain, Persia, and India. Long after the general adoption of Arabic numerals, this encyclopedic data continued to be recorded in the old Babylonian sexagesimal system, which used letters of the alphabet that stood for numbers. Greek astronomical manuscripts had also employed this system.

Figure 6.11
Ruznama, Ottoman, Paper and Wood, Turkey, Seventeenth Century

Almanacs similar to the one shown here, known as *ruzname*, were popular in Ottoman Turkey. This example mixes elements of religious practice, cosmology, local geography, and magic with mystical overtones. At the top of the scroll is a circle displaying the sacred Ka'ba at Mecca and identifying the major sites in the surrounding mosque compound. In addition, eight winds are identified by name, together with some mysterious, possibly magical, numbers. The purpose of the circle was to assist the user in summoning help from a hierarchy of legendary individuals. Beneath the circle are tables for converting between calendars and for finding the position of the sun on the ecliptic according to the date. The main group of tables is for regulating the times of prayer for the latitude of Istanbul. The entries are given in hours and minutes and are reckoned according to the peculiar Ottoman convention of figuring sunset at twelve o'clock. Under this system, still in use in some parts of the Muslim world, clocks must be reset at sunset every few days. In the margins of the tables is agricultural and astrological information for each of the months of the Syrian year. Tables similar to these were compiled for provincial capitals of the Ottoman empire from Algiers to Aleppo and San'a'.

ASTRONOMY

Figure 6.12
Astronomers at Work in the Observatory of Murad III (Takiyuddin in His Observatory at Galata), Illustration from *Shahinshahnama* (Book of the King of Kings), Vol. I, by Loqman, Turkey, Sixteenth Century

Islamic astronomical observation began in the eighth century and was accelerated by the demand for precise tables needed in preparing up-to-date calendars and horoscopes. The astronomers shown here, working in the sixteenth-century Istanbul observatory built by Murad III for a clockmaker who became an astrologer, are using a range of observational instruments that includes an astrolabe, an alidade (a rule used in determining direction) attached to a quadrant, and, at upper left, a diopter, a measuring instrument with viewing apertures.

Figure 6.13
The Underground Arc of the Great Observatory
at Samarkand, Uzbekistan

The long tradition of Muslim astronomical observation reached its zenith in great observatories such as those established in 1259 by the Mongol ruler Hulagu at Maragha, Azerbaijan, in Persia, and in 1420 by Sultan Ulugh Beg, Tamerlane's grandson, at Samarkand, in Uzbekistan, Central Asia. This is the graduated subterranean arc of the huge instrument (probably a sextant) used in taking readings at the Samarkand site. With a radius of about 130 feet, the trough supports the large arc, which is set in the plane of the terrestrial meridian. Light from celestial bodies crossing this plane comes through the opening, to fall on the graduated curving base, allowing altitude readings to be taken of the object casting the light.

ASTRONOMY

Figure 6.14a
Samrat Yantra (Main Sundial), Eighteenth-century
Astronomical Observatory, Jaipur, India

This and similar observatories at Delhi, Jaipur, and other Mughal cities were built by the Mughal ruler Jai Singh. Shown here is the observatory's largest instrument; it has a gnomon, the shade-casting marker, 90 feet high, flanked by shadow scales used to measure the changing shadows and thus determine the sun's exact position.

Figures 6.14b and 6.14c
Rasi Valaya and Jai Prakash Structures, Astronomical Observatory, Jaipur, India

In addition to the large instruments shown in the preceding figure, the observatory structures include smaller, zodiacally aligned sundials, as well as hemispherical bowls featuring graduated scales and openings through which observers inside the bowls can sight celestial bodies and record their passage.

ASTRONOMY

Figure 6.15
Astronomers Working with an Armillary Sphere, MS Illustration from
Shahinshahnama (Book of the King of Kings), Turkey, Sixteenth Century

The armillary sphere, an observational structure designed to represent the great celestial circles such as the equator, the horizon, the meridian, the tropical and polar circles, and the solar path, dates back to the Alexandrine Greeks of the second century AD. As they did with other ancient devices, the Muslims added refinements to the sphere. This illustration from a sixteenth-century Ottoman manuscript shows a giant device containing the fundamental circles; it is being used out-of-doors. The inner circle of the meridian ring could be rotated to provide measurements of solar altitudes and the angle of the sun's path.

[87]

Figures 6.16a and 6.16b
The Astrolabe

Perfected by scientists and master craftsmen working in the Islamic lands, this instrument, probably invented by Greeks in the second century BC, displays a mathematical model of the heavens. It can be manipulated to provide year-round celestial and timekeeping data, terrestrial measurements, and astrological information, thus solving many kinds of astronomical and timekeeping problems, including the determination of prayer times and establishment of the local direction of Mecca. The basic astrolabe usually has four major parts. A rotatable rete, the grid, bears pointers representing prominent stars and a circle representing the ecliptic, the sun's apparent path against the stars. Beneath the rete (see Fig. 6.16b) is a removable plate that fits into the mater, a flanged compartment, and bears markings representing the horizon for a specific latitude, the meridian, and altitude and azimuth circles. Most astrolabes have several plates for different latitudes. On the instrument's back is a movable alidade, a sighting bar, along with various engraved scales and markings used in astronomical and astrological calculation. Astrolabes are generally very compact, with diameters usually ranging between five and ten inches.

ASTRONOMY

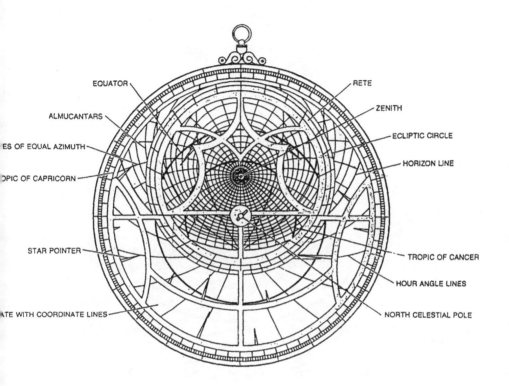

Figure 6.16a
Diagram Illustrating the Astrolabe

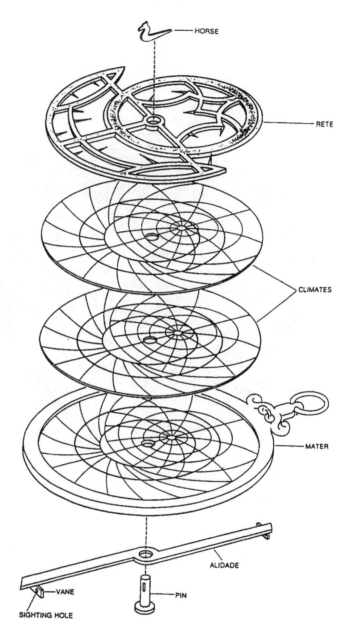

Figure 6.16b
The Parts of an Astrolabe

ASTRONOMY

Figure 6.17
A Twelfth-century Persian Astrolabe,
Signed Hamid ibn Mahmud al-Isfahani

Made, as were most astrolabes, of brass, this is a typical example of its time and place. It is inscribed on the back, in Kufic, a type of Arabic script, "Made by Hamid b. Mahmud al-Isfahani in the year 547" of the Islamic calendar.
It is the only known instrument by this craftsman.

Figures 6.18a and 6.18b
An Eighteenth-century Persian Astrolabe, Front and Back, Brass, made by 'Ali ibn Hassan Muhammad Khalil, decorated by 'Abd al-A'Imma, Isfahan

Bearing the signature of 'Abd al-A'Imma, the most famous of the Persian astrolabists, this exquisitely made instrument combines excellence in ornamentation with scientific accuracy. The rete fits over six plates, inscribed for twelve different latitudes, one for Mecca and the others for localities in Persia. The mater is inscribed with a geographical table displaying coordinates and *qiblas* of over forty localities, mostly in Persia. On the back is a circular grid for measuring celestial altitudes around the perimeter, a sine quadrant for performing trigonometric functions, and sets of curves for finding midday solar altitudes in different latitudes throughout the year, as well as altitudes for various localities when the sun is in the direction of Mecca. In addition, along with astrological information, the back contains a standard "shadow box" used with the alidade in reckoning the lengths of shadows from solar altitudes. By the time this astrolabe was made, most astronomical enterprise was directed toward astrology, and the level of Persian scientific activity had receded from the heights reached seven centuries earlier.
Probably only a court astrologer could have afforded such a splendid instrument as this one.

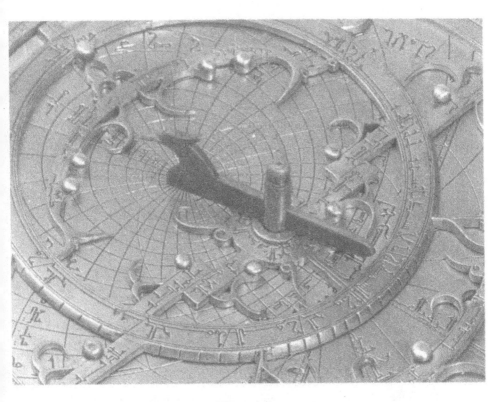

Figure 6.19
A Fourteenth-century Spanish Astrolabe—Detail of Rete, Brass,
Signed Ahmad ibn Husayn ibn Baso, Spain

This instrument is signed by Ahmad ibn Husayn ibn Baso, a well-known astronomer, as well as the *muwaqqit* of the Great Mosque of Granada when the Kingdom of Granada flourished under the Nasrids. The rotatable rete bears twenty-eight stars decorated with silver points.

Figures 6.20a and 6.20b
A Universal Astrolabe, Front and Back, Brass, Made by
Ahmad Ibn al-Sarraj, Syria, 1328–1329

This is the only known surviving example of an Islamic universal astrolabe. It was made in the fourteenth century by Ibn al-Sarraj, a prolific author of astronomical treatises who worked in Aleppo. The ordinary astrolabe is fitted with a group of plates serving a series of terrestrial latitudes; the universal astrolabe requires only one plate, for latitude zero, to achieve the same results. The instrument shown here is a doubly universal astrolabe: it incorporates not only its single plate and rete, or "spider," but also a series of plates each quadrant of which is marked for a specific latitude between the equator and the north pole. One of the plates consists of a complete set of terrestrial horizons. The astrolabe also bears on its back a universal trigonometric grid of highly sophisticated design, which can be used for solving numerically (using trigonometric formulae) all of the different astronomical problems which can be solved by using the front of the instrument as an analog computer. Usable in four different ways, the universal astrolabe was developed by the eleventh-century Toledan scholar al-Zarqali as a modification of an instrument devised by his contemporary, Ali ibn Khalaf. It has been described as the most ingenious astrolabe made during the entire medieval and Renaissance periods.

ASTRONOMY

Figure 6.21
A Fifteenth-century Spherical Astrolabe

Signed, "Work of Musa, year 885" (AD 1480–1481), this instrument, made of brass with damascened and laminated silver, has a diameter of slightly more than three inches, making it quite a bit smaller than most of the other, planispheric (flat plane) astrolabes shown in these pages. Spherical astrolabes were rare; this is the only example known. The names of the signs of the zodiac appear on the diagonal ecliptic circle. The rete, designed as a shell encasing the globe, has pointers for nineteen fixed stars. The globe is incised with lines representing the horizon, altitude lines, the azimuths, and other celestial markings.

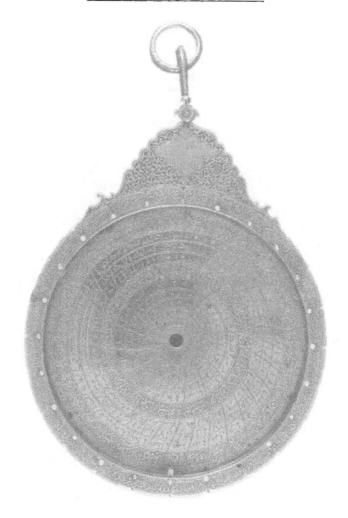

Figure 6.22
An Exquisitely Detailed Astrolabe Mater, Brass with Turquoises,
Persia, Seventeenth Century

Decorated with turquoise and signed "Decorated by the rich (in God), the servant, the beginner, the son of Muhammad Amin, Muhammad Mahdi al-Yazdi," this seventeenth-century Persian astrolabe is engraved in such extraordinarily fine detail that it gives credence to the story that some inhabitants of the Yazd-Isfahan region are genetically endowed with vastly superior eyesight. Such a faculty seems to have been possessed by this instrument's maker, who did not have access to a strong optical magnifier.

Figure 6.23
An Astrolabe with a Date Converter—Detail, Planispheric, Brass,
Morocco, Eighteenth Century

This unusually large (nearly ten inches in diameter), well-made eighteenth-century instrument bears the signature: "Praise to God! The maker of this astrolabe is the servant of his Lord, Muhammad b. Ahmad al-Battuti. May God pardon him and all Muslims!" The back contains an unequal hour diagram (shown here) with precise instructions for use with the date conversion scale. The instructions are not universally valid, but they worked for the Islamic calendar year (AH 1136) in which the astrolabe was made and for sixteen years before and after that date.

Figures 6.24a, 6.24b, 6.25, and 6.26
The Astrolabic Quadrant

The quadrant is a simplified astrolabe. The apparent daily rotation of the celestial sphere is simulated by the movement of a taut thread attached at the center of the instrument, with a bead being moved on the thread to a position corresponding to that of the sun or a given star, such positions being read from the quadrant's markings. The thread and bead thus take the place of the astrolabe's rete. Much easier to construct than an astrolabe, the astrolabic quadrant can be used to solve all of the standard problems of spherical astronomy for a particular latitude. Developed by Muslims in Egypt in the eleventh or twelfth century, it had, by the sixteenth century, replaced the astrolabe everywhere in the Muslim world except Persia and India. It should be noted that the examples illustrated in this section lack the requisite thread and bead.

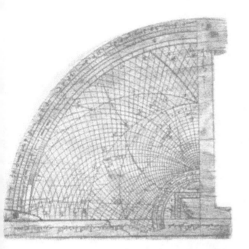 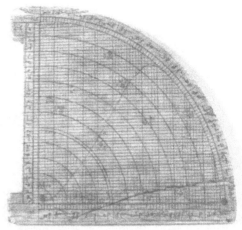

Figures 6.24a and 6.24b
Fourteenth-century Egyptian Quadrant, Ivory, Front and Back,
Egypt, AH 741 (AD 1340–1341)

This fine quadrant is unusual in that it is made of ivory instead of the usual brass or wood and serves two latitudes rather than one. The inner, standard set of markings on the front serves the latitude of Cairo; the outer, nonstandard set serves the latitude of Damascus. The back of the instrument bears a standard grid for use in solving trigonometric problems numerically. This quadrant also bears unusual markings devised by Ibn al-Sarraj, the fourteenth-century astronomer who made the universal astrolabe shown in Figures 6.20a and 6.20b.

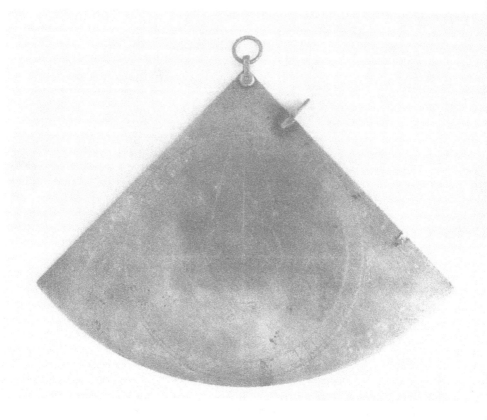

Figure 6.25
A Sixteenth-century Brass Quadrant from North Africa

This elegantly engraved brass quadrant bears a standard sinecal grid for performing trigonometric functions; the grid is the medieval equivalent to a slide rule. The back of the instrument (shown here) bears interesting markings that may be incomplete. The outer circle probably represents the celestial equator; the smaller circle is unmarked and serves no obvious function. The crescent is a stereographic projection of the ecliptic. Interpretation of the remaining elements—four curves distinguished by Arabic letters or symbols—must await the discovery of the secret of such medieval markings.

Figure 6.26
An Astronomer Observing a Meteor with a Quadrant, MS Illustration
from the *Nusratnamah*, Sixteenth Century

Knowing the coordinates of his own location, the astronomer lines up a celestial object with the sighting points on one edge of the quadrant. At that point, the location of the quadrant's string and movable bead in relation to the grid engraved on one of the instrument's sides indicates the celestial object's position in the heavens. This represents only one of several possible functions of this instrument, which, while much simpler than its predecessor, the astrolabe, could generally be used only in the one specific latitude for which its grid was designed.

Figure 6.27
Compass with a View of Mecca

Around the fourteenth century Muslims began to make a variety of instruments that combined a small sundial with a magnetic compass and a diagram or map indicating the direction of Mecca from a variety of cities. Ultimately this kind of instrument evolved into pocket-sized *qibla* indicators that showed the user within a large area how to determine the direction of Mecca. Ottoman instrument-makers, in particular, produced a multitude of illustrated instruments such as this one.

ASTRONOMY

Figure 6.28
A Seventeenth-century Celestial Globe, Brass, Persia or India, 1650

Inheriting the celestial globe from the ancient Greeks, medieval Muslims produced hundreds of them. Most were made of brass, and they were carefully, sometimes exquisitely, engraved with outlines of constellation figures, as well as numerous stars. Sometimes they featured the brighter stars by themselves or were marked with a grid of spherical coordinates. Often the stars consisted of silver inlay. Generally used as a teaching device, the celestial globe illustrates problems of spherical astronomy. The ecliptic is indicated against the background of stars beyond. The globe's axis is fixed in the plane of the meridian; however, its inclination to the horizon can be adjusted by rotating the globe so that the instrument displays the heavens with respect to any local horizon.

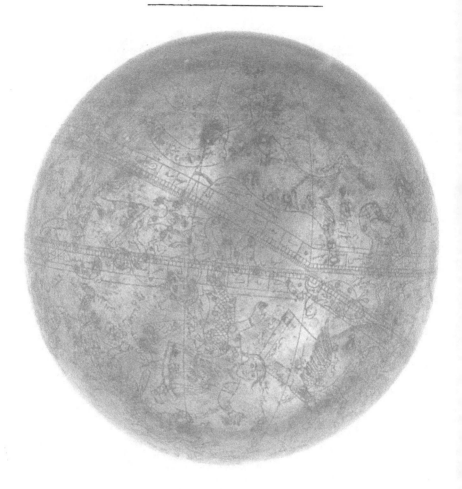

Figure 6.29
A Celestial Globe from Seventeenth-century India, Brass, Indo-Persian
(Lahore), Seventeenth Century

This splendid globe, nearly ten inches in diameter, is unsigned and undated but is thought to have been made at Lahore about 1620. Its markings are based on the star catalog published by Ulugh Beg in Samarkand a couple of hundred years earlier. The frame bearing a horizontal ring representing the local horizon has been lost. Some forty-eight constellations are depicted, as well as the ecliptic and celestial equator. Exquisitely wrought, this globe would have been treasured by any Indian ruler or teacher of astronomy.

Figures 6.30 and 6.31
The "Tusi Couple" and a Possible Descendant

Ptolemaic planetary theory accepted the principle of uniform circular motion but allowed the planets to move in epicycles, little circles, which themselves moved around deferents, larger circles. Despite their general allegiance to Ptolemy's cosmos, Muslim astronomers came to object to the way his epicyclic motions violated the principle of uniformity of motion firmly accepted by all astronomers, Ptolemy included. The result of their study was an important reform of planetary astronomy, effectively begun by al-Tusi in the thirteenth century. His famous "Tusi Couple," shown above at the left, illustrated a hypothetical model of epicyclic motion that involved a combination of motions each of which was uniform with respect to its own center. Developed more fully in the following century by Ibn al-Shatir, this model represented the most important Muslim achievement in planetary theory.

In recent years historians of astronomy have become interested in the marked similarity between the models of epicyclic motion proposed by al-Tusi and Ibn al-Shatir and one (shown at the right) proposed three centuries later by the great Polish astronomer Copernicus. It has never been established that Copernicus borrowed directly from his two Muslim predecessors, but this link between the Ptolemaic and Copernican concepts of the planetary system remains an intriguing possibility and is considered by some scholars to be a probability.

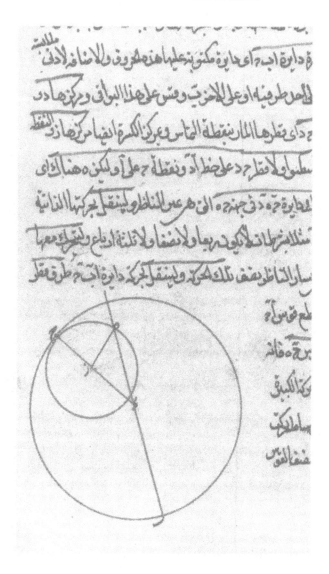

Figure 6.30
Diagram Illustrating the "Tusi Couple," MS Illustration from
Sharh al-Tadhkira, a Fifteenth-century Commentary by al-Birjandi on the
Thirteenth-century *Tadhkira* (Compendium of Astronomy)
by Nasir al-Din al-Tusi, Persia

larabimus. Interim uero quæret aliquis,
nodo pofsit illarum librationum æquali=
?i, cum à principio dictum fit, motum cęle
:x æqualibus ac circularibus cōpofitum.

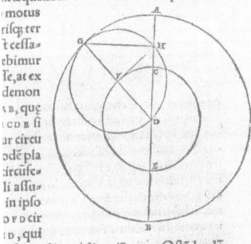

motus
rifcǫ ter
t ceſſa=
ebimur
le, at ex
demon
ιʙ, quę
ιcd ʙ fi
ɔr circu
ɔdē pla
:ircūfe=
li aſſu=
in ipſo
ο ꜰ ᴅ cir
ιᴅ, qui
ʜ ſigno, & agat dimetīes ᴅ ꜰ ɢ. Oſtēdendū
bus circulorū ɢ ʜ ᴅ & ᴄ ꜰ ʙ cōcurrētibus in=
ː rectam lineā ᴀ ʙ hinc inde reciprocādo re
ligat ʜ moueri in diuerfam partē, & duplo
: idē angulus, q̄ ſub ᴄᴅ ꜰ in cētro circuli ᴄ ꜰ ʜ

Figure 6.31
Diagram Illustrating Planetary Movement, MS Illustration from
De Revolutionibus by Nicolaus Copernicus, Nuremberg, 1543

7

Astrology
SCIENTIFIC NON-SCIENCE

*P*rimarily concerned with revealing the physical nature of earth's celestial environment, astronomers in the Islamic world were also interested in applying knowledge of the movements of heavenly bodies to the prediction of events affecting the lives of earth's people. Their astrology involved most of the same instruments and mathematical disciplines that were used in observational and theoretical astronomy. Muslim astrological observations and calculations demonstrate a scientific character; their interpretation, however, depended on the metaphysical procedures of divination to explain the changing configurations of the heavens and their meaning for daily life.

The sources of Muslim astrology reach as far back in history as those of Muslim mathematics or astronomy. Babylonians, the Sabians of Harran in northern Mesopotamia, Egyptians, Greeks, Indians, Persians, and Chinese all used observations of stars and planets to reveal the course of present and future happenings of every kind—political, military, environmental, and personal. The "rule of the heavens" has been accepted around the world for several millennia as all-powerful, to be disregarded only at the greatest risk to an individual, community, or dynasty. Humankind has always read the stars and will undoubtedly continue to do so.

Muslims inherited an ancient, honored, and rich tradition of astrological practice and, like virtually all societies previous to them, patronized astrologers and their practice at all levels. There was a double motive for the establishment of great observatories such as those at Maragha and Samarkand: the reigning caliph or prince felt obliged to equip his court or capital with the best possible facilities for plotting the motions of heavenly bodies, not only for purely astronomical purposes but also (occasionally primarily) for providing the royal astrologers with precise data on which to base their interpretations and predictions, so essential to political and military planning. Thus dynastic pride, superstition, royal vanity, and enthusiasm for science all helped to make astrology a permanent and vastly popular enterprise throughout the medieval world, Muslim and non-Muslim.

Astrology in the Islamic world was nurtured by the Muslims' preeminence in developing observational instruments of unprecedented capability, by their enhancements in mathematical calculation, and their increasingly sophisticated methods of astronomical inquiry and analysis. These factors provided medieval astrology with a notably scientific flavor, if not a notably rational character. Respectable support was available: just as Ptolemy's concept of celestial mechanics underlay much Muslim astronomical endeavor, his *Tetrabiblos*, a study of astrology, was a basic text for Muslim astrologers. Muslims produced a large body of astrological texts, especially between the ninth and fourteenth centuries.

As in medieval Europe, astrology in the Islamic lands was vastly popular among all classes. It also generated sharp opposition, especially among theologians and philosophers, such as al-Farabi and Ibn Sina. The Qur'an warns that no one in the heavens and earth knows the unseen except Allah (sura XXVII, verse 65). Religious authorities expressed their rigorous antagonism, even condemnation, by reminding the faithful that the fundamental Islamic concept of submission to the will of God rules out searching for predictions. Although al-Biruni, the Persian philosopher-astronomer, produced astrological manuals, he indicated that he had little confidence that the "decrees of the stars" belonged among the exact sciences. Muslim opposition to astrology was expressed at length in the *Muqaddima, An Introduction to History*, by the celebrated historian-philosopher Ibn Khaldun, born in North Africa in the early fourteenth century. Ibn Khaldun denied the stars' influence on the world below, described God as the only agent, and considered astrological achievement, even when considered rationally, as weak and harmful to mankind. Muslim theologians also opposed what came to be perceived as astrology's connection with predestination, Greek philosophy, and other "foreign sciences."

Despite disapproval by noted philosophers and men of letters, as well as by religious scholars, let alone religious law, medieval Muslim astrologers flourished, applying their skills to meet a wide variety of needs for prognostication, public and private. If the caliph was planning a new military campaign or the building of a new city, the royal astrologer was commanded to determine the most propitious time for attack or for laying the foundation stone. If the physician's patient needed surgery, the stars and planets would indicate the auspicious times for the operation (as well as the inauspicious ones). If the mariner was trying to find out whether or not his storm-tossed ship would make it to port, if the gravely ailing king demanded to know how long he had to live, and, later, if his mourning heir wanted to learn about his own prospects, the local or in-house astrologer was summoned to consult the heavens. The same practice was also followed at a more lowly and immediately practical level by a farming community when it was especially worried about the prospects for the coming harvest.

The procedures followed by medieval Muslim astrologers were generally the same as those that have been followed by astrologers everywhere since antiquity. The population of the heavens—sun, planets, and stars—were the core players in the astrological spectacle; their changing relationships provided the dramatic action that unfolded before the terrestrial observer. Various combinations, conjunctions, aspects, and dispositions of these bodies had to be determined and interpreted for any given moment. So did masculine and feminine signs, as well as assigned characteristics such as darkness, color, happiness, and their opposites, along with aspects, correspondences, alignments, and formations such as trines, sextiles, squares, and triplicities—a complex array of astrological elements.

The stage on which the awesome astrological drama took place—the vast celestial sphere perceived to stretch above earth—was blocked out by Muslim astrologers in areas or sectors defined by the horizon, the meridians, and the ecliptic—the path of the sun seen against the changing background of stars. Here was the scientifically charted (and automatically mobile!) scenery within which every astrological reading could be plotted and interpreted. Here the signs of the zodiac and their subdivisions played their parts; here were the twelve celestial mansions of the ecliptic, the essential stations for the planets involved in astrological prediction. It was the relative arrangement of these elements of the celestial sphere at a given time that was believed to determine what kind of life a newborn child would have or what would be the effects of a major public occurrence.

With all the factors at hand, the astrologer undertook one of several kinds of procedures: he could work out the most propitious time for a proposed action as noted above, or answer a client's question concerning the welfare of an absent relative, or predict a client's future on the basis of the configuration of the mansions in the sky that prevailed on the date and hour of that client's birth. Developing precision and versatility in collecting and manipulating their astronomical data, Islam's medieval astrologers not only influenced (for good or ill) the course of community and individual life but also provided philosophical and metaphysical symbols which stimulated a kind of cosmological reflection that, while often condemned, retained an appeal for generations of Muslims.

After the tenth century the increasing volume and intensity of opposition to astrology generated a gradual withdrawal of eminent astronomers and mathematicians from the ranks of those who participated in astrological activity. This in turn allowed astronomy to be more easily supported by religious leaders and more firmly and broadly engaged in its service to the practice of the Faith. The mosque, playing an increasing role as a patron of the exact sciences, was thus deeply involved in the Islamicization that reshaped the character of some of the most important ancient sciences. And this "naturalization," in turn, helped to enhance the quality and lengthen the life of the medieval Muslim effort in science, especially in astronomy.

Figure 7.1
The Astrolabe Goes to War: *Battle between Bahram Chubina and Khusrau Parwiz*, Painting (Miniature) Showing an Astrolabist in Action, Persia, Sixteenth Century

In the celebrated romantic saga *Khusrau and Shirin*, written by the twelfth-century Persian poet Nizami and based on a pre-Islamic legend, Khusrau, princely ruler of the Sasanian empire, must endure many trials before finally winning the hand of his love, the Armenian princess Shirin. A careful inspection of this scene of battle between Khusrau and his enemy Bahram Chubina will locate the prince's astrologer using an astrolabe to determine the outcome of the struggle. According to Nizami, everything turned out well, and Khusrau awoke the next morning to find Shirin at his side. Royal astrologers, in and outside of court observatories, provided astrological interpretations of observational data on demand. Medieval Islam's rulers more often than not made use of these interpretations in planning important political and military ventures.

ASTROLOGY

Figure 7.2
A Traditional Arab Horoscope, from an Eleventh-century Commentary by Ibn Ridwan on Ptolemy's *Tetrabiblos*

Ibn Ridwan was an eleventh-century Egyptian physician noted for his work in preventive medicine. He also encouraged astrological interpretations and predictions in connection with medical practice. This diagram accompanies Ibn Ridwan's lengthy text and displays the planets in a horoscope format that reflects the enormous debt owed by Muslim scientists—astronomers and astrologers, in particular—to Ptolemaic cosmology. The basic design of a square intersected by a cross is one of the most common formats in both ancient and medieval astrology and shows the dominating influence of a concept of elements—forces and qualities (air, fire, water, earth, and hot, cold, wet, dry)—that can be traced back to Babylonian times. Muslim horoscopes presented this concept in many styles; some are very simple and austere, others intricately and colorfully painted or gilded.

ASTROLOGY

Figure 7.3
A Thirteenth-century Islamic Geomantic Device,
Brass Inlaid with Silver, Egypt or Syria

Once loosely termed an "astrological computer," this unique, beautiful, and compact instrument, approximately thirteen inches long and eight inches high, is made of brass inlaid with silver, gold, and copper. It was designed to be used in the practice of geomancy, which, like astrology, is a form of divination—foretelling the future through contact with divine sources or omens and other signs, heavenly or earthly. The dial faces bear domino-like patterns that form the basic geomantic figures. These figures are associated with the twenty-eight mansions of the moon's monthly course through the heavens and could be deciphered to answer questions about the future, locate hidden treasure, or give advice concerning a variety of possible actions. *Islamic Geomancy and a Thirteenth-century Divinatory Device*, by Savage-Smith and Smith (see Works Consulted at the back of the present volume) describes in detail this intriguing instrument and its operation.

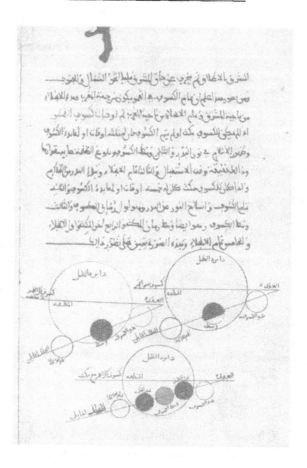

Figure 7.4
Astrology with Reservation: Diagrams Describing Lunar Eclipse, MS Illustration from *Kitab al-Tafhim fi Ma'rifat al-Tanjim* (Book of Instruction in the Elements of the Art of Astrology) by Abu Rayhan al-Biruni, Maghreb, Thirteenth Century

Despite Qur'anic admonition, astrology enjoyed widespread popularity and royal support during and following Islam's first expansion. This sponsorship also included the support of some eminent Muslim physicians who employed astrology in determining auspicious times for surgery. Even some of Islam's greatest philosopher-mathematicians, such as al-Kindi and al-Biruni, made use of astrological interpretations. Al-Biruni, however, in an astrological work that included this diagram of a lunar eclipse, professed little confidence in astrology as an exact science.

8

Geography

Islam's rapid expansion during its first few centuries was as much prompted by commercial incentives as it was by ideological and political ones. Muslims generated a vast exchange of commodities, achieving even greater volume and variety of cargo than had been carried by the galleys and wagons of imperial Greece and Rome. Many factors promoted the long-lasting supremacy of Islamic commerce. Trade had thrived in the Arab world for centuries. Muhammad had himself become a successful trader before the Islamic Revelation, thus commercial enterprise enjoyed an honorable professional standing among those who followed the Prophet's teachings. Trading with non-believers was not forbidden and was rarely interrupted for long periods, even during times of conflict.

The Islamic requirement of pilgrimage to Mecca, along with frequent journeying to other holy sites, was only part of a vast Muslim traffic. The rapid expansion, consolidation, and development of a dynamic new world civilization called for commerce on an unprecedented scale and accelerated the growth of convenient and well-serviced routes of travel in a region that was already a busy crossroads. Soon after the eighth century, the Mediterranean became virtually a Muslim highway. Heavily used sea and land routes eventually connected all the Muslim lands, extended around and through India to Southeast Asia and China, stretched north along the

Volga River to Scandinavia, and ran deep into Africa. Muslim raiding corsairs even reached Iceland. Linking commercial, cultural, and religious centers, this network carried generations of trading fleets and caravans bearing goods between all Muslim regions and some beyond, promoting a standard of living that was both remarkably uniform and more advanced than that generally known in Europe at the time.

This enormously productive traffic reflected more than traditional Arab and Bedouin practical skills in reliable geographic orientation and navigation, as well as in everyday business transaction. By the ninth century the Muslim empire embraced inhabitants from lands stretching from Spain to India, peoples with centuries of experience in long-distance as well as local commerce. Meeting the expanding needs of travel and trade as well as governance, Islamic geographers strove to clarify, extend, and fill in the picture of the known world they had inherited from ancient Babylonian, Iranian or Sasanian, and Greek sources, as well as from Jewish and Christian biblical texts and from China. Indeed, elements of some pre-Islamic geographical and cosmographical theories are preserved in the Qur'an. Muslims had gained a fair picture of those portions of Asia, Africa, and Europe that framed the Mediterranean basin and Asia Minor. Like most everyone else, they were for awhile daunted by the unknown world beyond: traditional wisdom reinforced the various ancient concepts of a great, presumably dark, and certainly dangerous ocean that extended in every direction beyond what was known. One such concept, for example, probably Persian or Sasanian, took truly fanciful form, presenting earth's landmass as a huge bird whose head was China, its wings south central Asia and India, its chest the Arab heartland, and its tail North Africa.

Ancient Greek and Roman concepts of earth's geography were maintained well into the Middle Ages. In the third century BC, the Greek astronomer Eratosthenes used lines of latitude to demarcate pairs of frigid and temperate zones, with one torrid zone sandwiched in between; his estimate of the earth's circumference came fairly near the figure generally accepted today. By the second century, when Ptolemy plotted earth's regions, their relative positions, and their distances from one another, meridians and parallels were providing a more accurate framework in which to refine the picture of the globe's land and ocean masses. It was the Ptolemaic view that formed the basis of Muslim cartography, as opposed to the more

[118]

fanciful and idealistic charts produced by medieval European mapmakers. Superimposed on this view were various Persian and Indian concepts, generally dealing with the division of the world into seven or nine regions; the traditional Greek partition involved nine climes. The notion that the earth's great Encircling Ocean fed the two known bodies of water (the Mediterranean Sea and the Indian Ocean) separated by the Isthmus of Suez was maintained almost to the end of medieval times.

By the tenth century, Muslim geographers were infusing Arab geography with a more specifically Islamic character, drawing increasingly on Qur'anic sources and on the *hadith*. A group known as the Balkhi School, including the notable geographers Abu Ishaq al-Istakhri and Ibn Hawqal, divided the lands of Islam into twenty regions, leaving the non-Islamic lands in a separate category. By this time the Islamic cartographic view of its lands was beginning to look more like our view today.

The discovery of chronicles recorded in pre-Islamic times by Egyptian, Phoenician, Cretan, and Greek geographers and mariners stimulated Muslim enthusiasm for investigating the far reaches of their world and beyond. Knowledge of these chronicles, which contained data and travelwise lore about remote regions, spurred Muslim geographers to record their own findings about the wider world. Sturdier, more resilient vessels were built, safer courses charted, additional way stations established. The growing feasibility of travel and the lure of exotic wonders described with awe in Muslim travel journals promoted further geographic inquiry, exploration, and mapping. At some point late in the eleventh century Muslims obtained from China a device, the compass, that utilized a magnetic needle to indicate direction. Within a century it was familiar to both Muslim and Christian mariners, who improved its dependability and accuracy.

Much of the contemporary geographical knowledge gathered by Muslims was set down by commercial travelers and pilgrims in detailed accounts of the lands, peoples, and customs they encountered on their voyages within and beyond the boundaries of Islamic territory. Officials serving at caliphal outposts, merchants traveling with great caravans, trading-ship passengers, scholars on their way to distant educational centers, pilgrims—many kept diaries, journals, or encyclopedic descriptions of their journeys and destinations. The tenth-century geographer, historian, traveler, chronicler, and philosopher from Baghdad, al-Mas'udi, produced *Muruj al-dhahab* (The Meadows of Gold and Quarries of Jewels), which describes places he visited and contains historical accounts of the ancient world, portraits of European subjects such as the French royalty, and comments on everything from monuments and customs to medicine and the nature of the soul. Al-Mas'udi took a dim view of Northern Europeans: he found them humorless, gross, and dull. Three centuries later the distinguished geographer Yaqut expressed a favorable impression of Rome. Interest in the ways of Europeans was not shared by great numbers of Muslims, however, until Western Europe's post-Renaissance achievements in science and technology—particularly military technology—began to be known and perceived as threatening.

The fourteenth-century Moroccan explorer Ibn Battuta spent his life traveling from North Africa to China and Southeast Asia, covering most of the lands in between. His accounts of what he saw and learned provided the literature of travel with some of the most objective, perceptive, and sophisticated observations ever made by a world traveler. Further and greater knowledge of lands and peoples within Muslim boundaries was provided in the same century by the North African historian Ibn Khaldun, sometimes called the father of sociology as well as of history.

The earliest Islamic atlases usually consisted of diagrammatic itineraries that showed towns and connecting roads in a style not unlike that found in today's fold-out automobile touring guides. The tenth-century maps prepared by al-Istakhri provide good examples of this type of cartography (Fig. 8.3). Probably the high point of Muslim mapmaking was achieved in twelfth-century Sicily, at the court of the Norman (and Christian) King Roger II, patron of the celebrated geographer al-Idrisi. At the king's request, al-Idrisi made a large silver relief map of the world and then reproduced its details on seventy-one separate maps, accompanying them with a descriptive treatise entitled *Kitab al-Rujari* (The Book of Roger). Al-Idrisi included much information, especially about Europe, that had never before been set down.

Al-Idrisi's works contributed greatly to the geographical education of Western Europeans, who were soon to embark on an age of exploration that would take them to all the continents of the globe. Muslim geographers played their part in this epic adventure: on his trip around Africa at the end of the fourteenth century the Portuguese explorer Vasco da Gama employed an Arab pilot who wrote a guide for mariners crossing the Indian Ocean. In the next century a Turkish Ottoman cartographer, Piri Re'is, produced an elegant and detailed atlas of the Mediterranean, which showed some influences of Western cartography. Interaction between the sciences of Islam and the West was underway in both directions, but the intensity with which the West utilized the knowledge it gained did not begin to be matched by similar Islamic effort until the twentieth century.

Figure 8.1
An Astrolabist Comes to the Aid of Noah's Ark: *Noah's Ark*, Painting Showing Mariner Using an Astrolabe, India, Seventeenth Century

Although the maritime scene depicted in this unusual Mughal painting of Noah's Ark portrays rather violent winds and seas, in addition to a just-in-time rescue of an overboard passenger from a sea beast's jaws, it is probable that the astrolabist shown at work atop the ship's stern at the left is trying to check the vessel's storm-tossed course rather than to foretell the outcome of the voyage. Noah was a popular figure in Qur'anic text and Muslim legend, and the outcome of the journey was well-known.

Figure 8.2
A Ship Crossing the Persian Gulf, MS Illustration from *Maqamat* (Assemblies) of al-Hariri, Iraq, Thirteenth Century

Following the Islamic conquests in the seventh and eighth centuries, Muslim commercial expansion brought both wealth and a sense of Islamic unity not only to Muslim territories east and west, but also to territories that extended east beyond India, north to Scandinavia, and south into Central Africa.

This illustration is one of many contained in a famous thirteenth-century edition of the *Maqamat* (Assemblies), written by the eleventh-century writer al-Hariri, who was born in Basra, south Iraq. This collection of entertaining and instructive stories about a wandering hero is illustrated by manuscript paintings that provide a uniquely detailed record of many aspects of daily life in medieval Islamic lands.

GEOGRAPHY

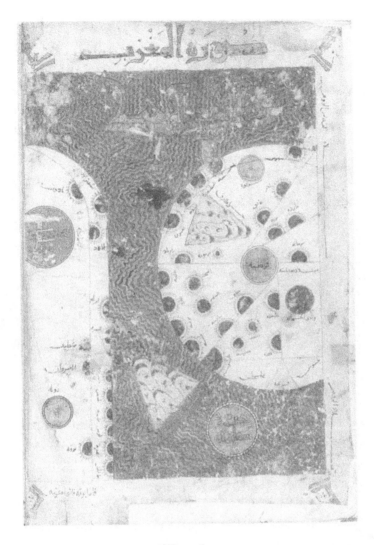

Figure 8.3
Map of Spain and North Africa, according to Fifteenth-century Copy
of Tenth-century Map by al-Istakhri, from *Monumenta Cartographica*
by Yusuf Kamal

In this fifteenth-century version of al-Istakhri's tenth-century map, North Africa appears on the left and Spain on the right. The circle at center right represents Córdoba, the great political, cultural, and commercial capital of *al-Andalus* (Andalusia).

Figure 8.4
Map of Turkestan from *Al-Masalik w'al Mamalik* (The Travel Routes of the Dominions) by Abu Ishaq al-Istakhri, Persia, Seventeenth Century

This seventeenth-century copy of a map originally included in al-Istakhri's tenth-century atlas, *Al-Masalik w'al Mamalik*, displays a typical schematic representation of cities and towns and the routes that connected them.

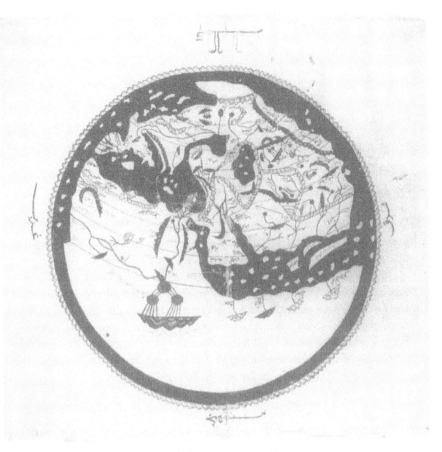

Figure 8.5
Al-Idrisi's Map of the World, MS Illustration Copied from
Kitab al-Rujari, al-Idrisi, Sicily, Twelfth Century

In the twelfth century the Moroccan geographer-cartographer al-Idrisi completed his encyclopedic work, *Kitab al-Rujari* (The Book of Roger), dedicated to his Christian patron, Roger II of Sicily. Al-Idrisi's world map, shown here upside-down, presents a recognizable outline of the Mediterranean basin and Central Asia. It displays the Ptolemaic system of dividing the inhabited world into seven climes; these were bounded by circles north of and parallel to the equator and were divided longitudinally into ten sections. This particular map was probably based on a famous representation made by Idrisi in solid silver. Medieval Muslim maps, as well as Muslim navigational instruments, guided many mariners in Europe's Age of Exploration.

Figure 8.6
Map of the New World (the Americas) and West Africa by Piri Re'is,
Turkish, Sixteenth Century

This celebrated map, prepared on parchment in 1513 by Piri Re'is, a captain in the imperial Ottoman navy, shows the Atlantic shores of Europe and Africa, as well as those of Central and South America. The cartographer consulted many earlier maps, some of them from Hellenistic Greece, some Muslim, some Portuguese, and one thought to have been made by Christopher Columbus. This map has given rise to considerable controversy in connection with the sources used in its preparation and, in particular, with its display of regions long thought to be unknown at the time the map was drawn. What is certain is that here is an extraordinarily detailed cartographic achievement that reveals many aspects of sixteenth-century geography, geology, wildlife, legend, and mythology, as well as maritime commerce. It is also an early record of the sharing of vital knowledge between the Muslim and Christian worlds.

GEOGRAPHY

The best gift from Allah to mankind is good health. Everyone should reach that goal by preserving it for now and the future.

Medicine

One of the *hadith* attributed to Muhammad, the statement above reflects the serious and sustained attention devoted by Islamic society to the life sciences. The heritage of knowledge absorbed by medieval Muslim physicians was indeed rich. It embraced the achievements of classical Greece and Rome as well as those learned from Syriac, Persian, and Indian sources.

The Greeks conceived of four constitutional humors (melancholic, sanguineous, choleric, and phlegmatic), along with three functional fluids (arterial and venous blood and nervous fluid) and the spirits (vital, natural, and animal) that controlled the fluids. All these elements, in the proper balance, were essential for good health. They were spelled out in the writings of Hippocrates and, above all, in the basic teachings and elaborate theories of Galen, which dominated the formation of Muslim medical theory and practice.

Greco-Roman medicine had come to define illness in general as a natural phenomenon within the humoral framework. Muslim physicians reasoned that the humors were profoundly influenced by life's strains and stresses. It was thought that illness could be countered by utilizing the patient's own systemic resources in helping along the processes of curing or healing. This kind of thinking was far removed from the notions of "evil spirits" or forces of magic so often connected with disease and treatment in the medieval West and elsewhere. Furthermore, such an approach was in certain ways implicit in Islam's religious thought, and Muslim physicians applied it to every aspect of their own theory and practice.

One of the Muslims' greatest sources of practical information, used by both physicians and pharmacists, was *De Materia Medica*, written in the first century AD by Dioscorides, a Greek surgeon on duty with the Roman army. This work described in detail a remarkably broad range of drugs and the plants from which they could be extracted. Interest in medicinal herbs had been a feature of Indian medicine centuries before the birth of Islam, and the Hindus also stressed the importance of a proper balance between the different physical elements that controlled the vital processes of life.

As was noted above, by the sixth century AD the Persian city of Gondeshapur had become an intellectual and medical center. It was also a haven for many of the religious and political refugees who had fled from the Eastern Roman Empire between 489 and 529, when Christian authorities shut down pagan educational institutions, including the academy in Athens established by Plato eight centuries before. Scholars brought to this city translations of philosophical and scientific manuscripts, including those dealing with medical subjects. Families of medical specialists, many of them belonging to Syriac Nestorian religious denominations, developed into virtual dynasties that included the first professional medical teachers to train Muslim practitioners after the Arabs captured the city in 638 AD. Eventually many descendents and successors of these medical specialists moved to Damascus and then to Baghdad when these cities, in turn, became the capitals of the Umayyad and Abbasid dynasties.

This extraordinary and diverse core of gifted scholars, teachers, and practitioners did much to secure the foundation for broad and unprecedented Muslim achievements in virtually all aspects of the healing arts. With considerable support provided by caliphs such as al-Ma'mun in Baghdad, the treasury of Greek texts was soon made available in Arabic translation. Much of this production was due to the prodigious efforts of the celebrated Hunayn ibn Ishaq, who together with his school of translators rendered into Arabic most of the works of the great Greek medical triumvirate (Hippocrates, Dioscorides, and Galen), thus providing virtually all Islam's early medical students with their basic reference texts. At a very early stage of Islam's development there were in place the elements needed for unparalleled advancement in all the medical arts: libraries had proliferated, translation centers had spurred the dissemination of the wisdom collected from the past, and the hospital as an institution was being developed in revolutionary ways that would shape the course of health sciences and health care right down to modern times.

Hospitals

The Muslims' organizational talents, along with their special clinical and surgical skills, were applied with particular success in the development of great hospitals in the major cities of medieval Islam. Both in size and professional expertise, these institutions far surpassed virtually all medical institutions known in ancient times or outside of Muslim lands during the Middle Ages.

In medieval Europe most hospitals were associated with religious orders, monasteries in particular. Beyond utilizing pharmacies and gardens of medical plants, medical care, while compassionate, was oriented more to filling the patient's spiritual needs than to treating bodily disease and injury. Thoughtful attention was given to mental illness in the medieval West, but some of the less severe manifestations of psychosis were frowned on by the Church as symptoms of laziness.

Hospitals as we know them today were first developed in Islamic lands more than a thousand years ago. The first and most elaborate was built in the eighth century during the reign of Caliph al-Rashid (the caliph of the "Thousand Nights and One Night"). It was soon followed by similar institutions in one city after another throughout the Islamic empire. They were paid for and maintained by endowments from caliphs, other rulers, philanthropists, and religious foundations, and were managed and operated by highly educated and professional staffs. The conquering Arabs had the wisdom to preserve the preeminent medical institutions they found already established in North Africa and Asia, such as the famous Persian academy and hospital said to have been founded in the sixth century at Gondeshapur. Before long, dozens of great Muslim hospitals flourished between Asia Minor and the Maghreb (the Muslims' name for their coastal territories in Northwest Africa).

The typical Islamic hospital complex, as it existed in Cairo, Baghdad, Damascus, and later in many Turkish cities, was a remarkably resourceful enterprise. In addition to medical facilities, it often contained, or was attached to, a mosque, a *madrasa*, and sometimes a mausoleum honoring the founder, whether caliph, sultan, prince, or other local eminence. These monumental establishments, transplanted intact to virtually any Western metropolis today, would hardly seem a thousand years behind the times; perhaps they might seem somewhat mid-Victorian in character, with an "Eastern" ambience (Fig. 9.2c).

The medieval Muslim hospital's organization and layout were considerably advanced for their time. Separate wards were provided for male and female patients. Special wards were maintained for internal diseases, ophthalmic disorders, and orthopedic cases, as well as for other surgical patients, the mentally ill, and patients with contagious diseases. Extensive training and pharmacological facilities were standard. All great hospitals were host to physicians who came from all parts of the Muslim world to function as resident administrators, specialists, practitioners, and visiting teachers. Apprenticeships were offered to deserving students. Outreach was considerable, and traveling clinics and dispensaries provided wide areas with professional treatment and care. Military hospitals were organized to accompany armies on the move.

There is interesting evidence concerning medical ethics and the doctor-patient relationship in medieval Islamic lands. Contemporary texts indicate that these matters were handled in sophisticated and advanced ways. In his days of training and study the physician received all kinds of advice. One of the first Arabic texts to treat the subject of medical ethics, al-Ruhawi's *Adab al-Tabib* (The Physician's Code of Ethics), written in the ninth century, had much to tell the intern, in particular. He was counseled to walk and talk becomingly, to be modest, virtuous, kind, and merciful, to avoid money-grubbing, slander, and addiction to wines and drugs. He was also advised to avoid losing his temper and to answer patiently when asked many questions by the patient or the patient's family. The physician was urged to render his best professional services to rich and poor alike. Moreover, he was urged to give consent courteously if a client requested a second opinion, and, further, if that second opinion should run counter to his own, he was expected to explain both points of view to the patient.

In connection with visits to the patient, even the conduct of proper sickroom conversation was covered by Islamic religious law, which reinforced the tradition of humanistic personal consideration inherited from Hippocrates.

Physicians were not restricted to tried-and-true methods of treatment and could experiment freely. Their write-ups of their cases, including their conclusions as to the success of any experimental procedures, could be read by anyone. The training of medical personnel seems to have been remarkably thorough, utilizing celebrated teachers, who attracted students from great distances, as well as a rigorous system of examinations by *muhtasib*, authorities with the rank of judge or legal scholar. This testing determined which applicants would get the required and prized license or certificate to practice professionally.

The Great Physicians

The pantheon of outstanding Muslim physicians and medical specialists is large, and at least two of its members, al-Razi and Ibn Sina, must rank among the greatest physicians of all time. Al-Razi, known in the West by the Latin name Rhazes, was born in the Persian city of Rayy in the ninth century. He was an alchemist in his youth, subsequently gaining sufficient experience in a variety of medical activities to become celebrated throughout Western Asia, where he attracted throngs of students and patients. Like most other major Muslim intellectual and scientific figures, al-Razi was a polymath: while half of his nearly two hundred major written works and short treatises deal with medicine, the rest cover a wide range of subjects, including theology, philosophy, mathematics, astronomy, the natural sciences, and alchemy. Titles of some of his texts indicate that he was something of a practical psychologist, let alone a droll realist: *On the Fact that Even Skillful Physicians Cannot Heal All Diseases* and *Why People Prefer Quacks and Charlatans to Skilled Physicians* are among such works.

One of al-Razi's most famous texts, *On Smallpox and Measles*, translated into Latin and then English and other Western languages, went through forty editions between the fifteenth and nineteenth centuries. Al-Razi was probably the greatest clinician of the Middle Ages, and his most important work, *Al-Hawi* (Comprehensive Book), is one of the most extensive medical texts written by a doctor prior to the nineteenth century. Its twenty-three volumes constitute an encyclopedia of Greek, pre-Islamic Near Eastern, Syriac, Indian, and Arabic medical knowledge. The text may also reflect some of the author's contacts with Chinese medicine. Al-Razi's general method in this work involves describing the findings of all previous authors and then expressing his own opinions on the matter, usually beginning with the phrase "but I say . . ." or "for me . . . ." His texts covered a multitude of medical subjects, including surgery, clinical medicine, diseases of the skin and the joints, diet, and hygiene. His was a practical and rational mind, independent and critical of the tendency to thoughtlessly follow tradition, whether secular or religious.

Another towering figure, Ibn Sina, known in the West as Avicenna, was born in the tenth century in the Central Asian region of Bukhara. He was as learned in diverse subjects as was al-Razi. While celebrated as a philosopher, he also came to be considered the greatest Muslim writer on medicine. His encyclopedic *Al-Qanun* (The Canon) embraced all Greek medical knowledge together with Arabic interpretations and contributions. Divided into five books, *Al-Qanun* covers general medical topics such as anatomy, diseases, hygiene, and purges, simple therapeutic substances, disorders of the arm and leg, general conditions such as fevers, and tumors, fractures, even cosmetic problems. It also covers compound and other special medicaments, along with measuring instruments. A masterpiece of systematic organization, Ibn Sina's *Al-Qanun*, together with al-Razi's work, was used as the basic text in Europe's medical schools almost until the beginning of modern times. Ultimately it became probably the most used of all medieval medical references.

Believing medicine to be the art of removing impediments to the normal functioning of nature, Ibn Sina extended this concept to the study of human behavior and thus may be considered a pioneer in psychology. As his *Al-Qanun* demonstrates, he probed many aspects of human welfare that we often assume are purely modern concerns, such as the environment, cancer treatment, and psychotherapy; he even offered advice for those who were suffering the pangs of love! He perceived a close connection between emotional and physical states. And he has been quoted as being convinced that of all exercises for health, "music is better." This remarkable physician-sage inevitably received many honors and frequently held high office. He was proud, iconoclastic, and elitist; he was sometimes fiercely opposed, even imprisoned in times of political turmoil, and he saw his works ridiculed, discarded, and banned. Possibly he never imagined how profoundly his knowledge and wisdom would shape human welfare in the centuries after his death in 1057. His grave, in Hamadan, in present-day Iran, was ultimately honored by an impressive monument.

By the twelfth century, Muslim physicians had produced an enormous library of works: encyclopedias, medical biographies, texts on specialties such as ophthalmology, and guides to medical practice as well as to practitioners. With ample opportunities for observing the ravages caused by smallpox, cholera, and bubonic plague, physicians such as Ibn al-Khatib, a noted medical pioneer in fourteenth-century Granada, Andalusia, pursued his studies of epidemics in terms of contagion, a concept almost absent from medieval European medical writings. Among other Muslim medical developments was the growing critical examination of the works of Galen, the Greco-Roman physician of the second century AD; his thought had long influenced the character and course of Islamic medicine.

In the twelfth and thirteenth centuries a number of eminent figures, including Ibn Rushd, the great Muslim philosopher-physician from Muslim Spain, and Maimonides, the Jewish polymath (philosopher, mathematician, and astronomer, as well as physician) who became court physician to King Saladin, founder of the Ayyubid dynasty in Egypt, Syria, and parts of Arabia, ventured notably far beyond the boundaries of their intellectual inheritance. Maimonides' Egyptian contemporary, Ibn al-Nafis, put forth a new and challenging theory about the secondary, or lesser (pulmonary), circulation of the blood between the heart and the lungs. His work was ignored from the thirteenth century until its rediscovery in the twentieth. Meanwhile, in the West it was not until the seventeenth century that William Harvey's scholarly experiments finally revealed the blood's greater, or complete, circulation, one of the most important findings in the history of human physiology.

Medieval Muslim physicians added significantly to the knowledge of anatomy and physiology. For example, Hunayn ibn Ishaq's ninth-century treatises on the eye included remarkably accurate anatomical diagrams, the first of their kind in this field. Studies such as a fifteenth-century Persian manuscript, *Tashrih al-badan* (Anatomy of the Body), by Mansur ibn

Muhammad ibn al-Faqih Ilyas (Fig. 9.3), provided the most comprehensive diagrams yet seen of the body's structural, circulatory, and nervous systems.

Muslim achievements in surgery are all the more remarkable in view of the general religious disapproval of dissection of the human body. The most important surgical texts were written by a tenth- to eleventh-century Andalusian, Abu'l-Qasim al-Zahrawi, known in the West as Abulcasis. His *Kitab al-Tasrif* (Book of Concessions), a medical encyclopedia, contained three major surgical treatises which, translated into Latin, were used in Muslim and European medical schools for several centuries. The texts included al-Zahrawi's designs for about two hundred instruments, in addition to original and useful observations concerning topics ranging from surgical manipulations and technologies to cauterization, treatment of wounds, obstetrics, fracture, dislocation, paralysis, artificial teeth, and mouth hygiene. Al-Zahrawi's twelfth-century fellow-Andalusian, Ibn Zuhr, known as Avenzoar, was another of the great physicians who emerged during this period, the most brilliant in the history of Muslim culture in Spain. His works, especially those on anatomy, had considerable influence on medical practice in medieval Europe.

One of the oldest and most traditional occupations in world medicine has been that of the barber-surgeon, an important person in communities on all continents, a handy medical practitioner of tooth extraction and other non-major operations, as well as a beard-trimmer and hair-cutter. In Islam more than in any other medieval civilization, such a figure was only the most local participant in what rapidly developed into unprecedented regional networks of medical care for all. Like the neighborhood druggist in every medieval Muslim community, the barber was required by the local hospital to be periodically inspected and examined in order to be allowed to continue his practice legally.

Pharmacology and Pharmacy

Islam's achievement in medical research and treatment included notable progress in pharmacology, which may be defined as the scientific investigation of the composition, dosages, uses, and therapeutic effects of simple and compound drugs. Translations of Dioscorides' *De Materia Medica*, together with knowledge inherited from Syria, Persia, India, and the Far East, formed the basis for much innovation accomplished by eminent scholars and physicians in this field, in which Muslims remained unsurpassed until the seventeenth century.

To the store of medicines known since ancient times, Muslims contributed many items, including anise, cinnamon, cloves, camphor, myrrh, sulphur, and mercury. They were among the first to develop and perfect the manufacture of syrups and juleps, and they established the first apothecary shops. Arab and Persian treatises on drugs and compound remedies became famous throughout the Christian and Islamic worlds. One work in particular, Ibn al-Baytar's thirteenth-century *Al-Jamiʿ fi al-Tibb* (Collection of Simple Diets and Drugs), was based on actual plants gathered along the length of the Mediterranean coast between Spain and Syria. The author not only described more than a thousand different items but also compared them with those already recorded by more than a hundred scientists in previous eras and in regions ranging from Greece and Asia Minor all the way to China. His was the greatest Arab text dealing with medicinal botany; it surpassed that of Dioscorides and remained in use until the European Renaissance.

Pharmacy, specifically the art of preparing and dispensing drugs, first emerged as an independent profession in the Islamic lands during the late eighth century. It involved numerous practitioners, including herbalists, collectors and sellers of medicinal herbs and spices, manufacturers and sellers of syrups, cosmetics, and aromatic waters, field-trained apothecaries, and highly trained pharmacist-authors. The Muslim influence on later Western advances in pharmacology and pharmacy was considerable, and is reflected in the renewed interest in natural medicine that has become a popular feature of health education today.

Figure 9.1
Islam's Medical Inheritance: Portraits of Nine Greek Physicians, MS Illustration from *Kitab al-Tiryaq* (Book of Antidotes), based on the works of Galen, Probably Iraq, Thirteenth Century

Ninth-century Arabic translations of Greek texts by Hippocrates, Paulus of Aegineta, and Galen provided much of the basis of early Islamic medical education. This title page of a celebrated thirteenth-century copy of the *Kitab al-Tiryaq* presents portraits of nine Greek physicians. By the eleventh century the roster of illustrious Muslim physicians included two of the greatest names in all medical history, Ibn Sina and al-Razi, known in the West as Avicenna and Rhazes.

MEDICINE

Figures 9.2a, 9.2b, and 9.2c
Late Medieval Muslim Hospitals

Hospitals as we know them today were first developed in the Islamic world in the eighth century. They were supported by endowment from caliphs and religious foundations and managed by professional lay staffs. The photographs and plan shown on the following two pages illustrate the type of public complex that flourished especially in Mamluk Egypt and Ottoman Turkey. Vaccination against smallpox was introduced to Europe from Turkey in the eighteenth century; perhaps it was developed in facilities such as these.

Figure 9.2a
The Hospital at Divrigi, Turkey, Built in the Thirteenth Century

Figure 9.2b
The Hospital Complex of Beyazit II at Edirne, Turkey,
Built in the Fifteenth Century

MEDICINE

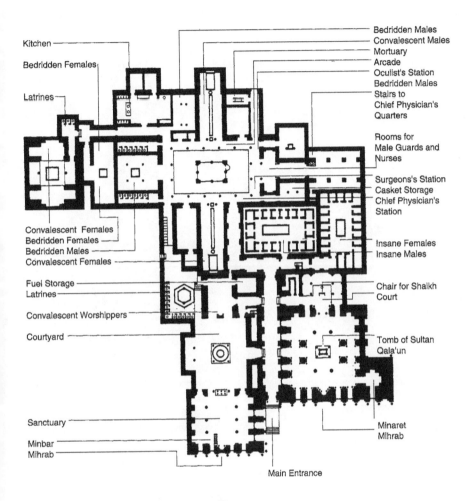

Figure 9.2c
Plan of the Hospital of Qalaoun, Cairo, Built in the Thirteenth Century,
Nineteenth-century Book Illustration

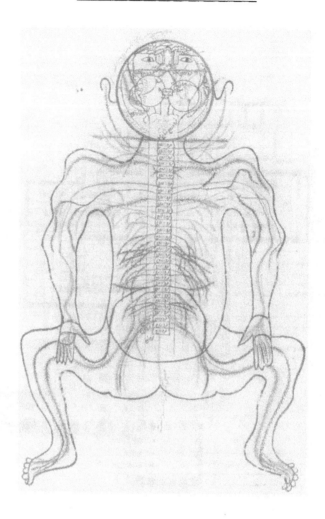

Figure 9.3
Diagram of the Human Nervous System, MS Illustration from
Tashrih al-badan (Anatomy of the Body) by
Mansur ibn Muhammad ibn al-Faqih Ilyas, Persia, Fifteenth Century

Medieval Muslim physicians added significantly to our knowledge of anatomy and physiology. Diagrams such as this fifteenth-century example reveal considerable understanding of the body's vital processes. Muslim achievements included a new theory about the secondary, or lesser, circulation of the blood (between the heart and the lungs) that remained generally ignored until its rediscovery in our own time.

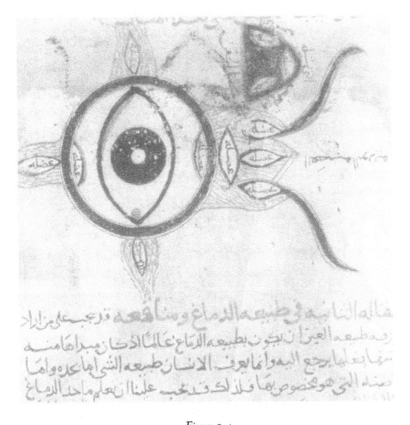

Figure 9.4
Diagram of the Eye, MS Illustration from *Kitab al-Ashr Maqalat fī l'Ayn*
(Book of Ten Treatises on the Eye) by Hunayn ibn Ishaq,
Egypt, Thirteenth-century Copy

Hunayn ibn Ishaq's ten ninth-century treatises on the anatomy and physiology of the eye, containing diagrams similar to the one illustrated here, provided the first comprehensive and systematic Arabic text on the subject. Although not as original in its exposition of a theory of vision as was Ibn al-Haytham's later text on optics (see Chapter 12 and Fig. 12.1a), Hunayn's remarkably accurate presentation of the eye's anatomy helped to establish his work as a standard educational and professional reference for centuries; it had a profound influence on the development of ophthalmology throughout both Islam and Europe. A scholar versed in many disciplines, Hunayn, together with his famous school of translators, rendered into Arabic all of the most important Greek medical treatises, as well as the writings of Aristotle.

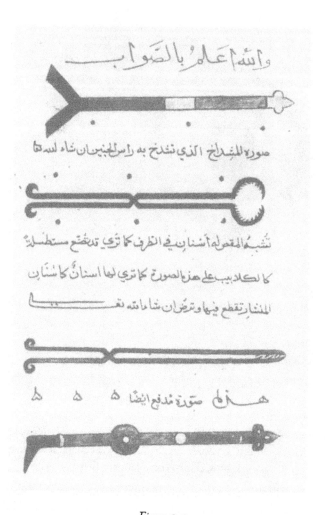

Figure 9.5
Medieval Muslim Surgical Instruments, MS Illustration from *Kitab al-Tasrif* (Encyclopedia of Medicine) by Abu'l-Qasim al-Zahrawi, Fifteenth-century Copy of an Eleventh-century MS Written in Spain

Important advances in surgical knowledge were made by the Andalusian al-Zahrawi, known in the West as Abulcasis, whose tenth- and eleventh-century texts included his designs for more than two hundred instruments, as well as a wealth of new observations concerning obstetrics, fracture, dislocation, mouth hygiene, and other surgical topics. His work guided students in medieval Europe.

MEDICINE

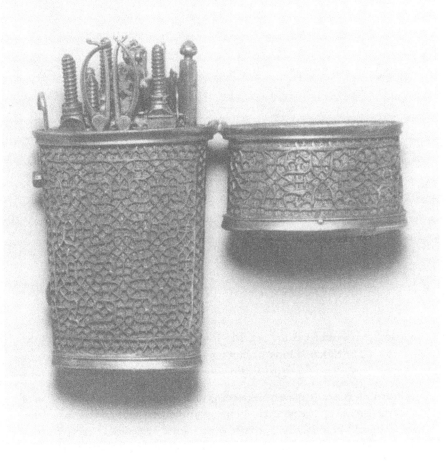

Figure 9.6
Eighteenth-century Persian Barber's Pocket Kit

The case with its surgical instruments includes a razor, pen, knife, scissors, file, powder spoon, and saw. For centuries the barber-surgeon, as he was known, was an important figure in both Muslim and non-Muslim communities, being qualified to perform minor operations, including tooth extraction, as well as traditional barber functions. This kit measures about 4 inches in height and 2 inches in width.

Figure 9.7
Dislocated Shoulder Being Set, MS Illustration from *Jarrahiyat al-Khaniyah* (The Sultan's Book of Surgery) by Sharaf al-Din ibn 'Ali, Turkey, Fifteenth Century

Whether the patient required the resetting of a dislocated shoulder, extraction of a tooth, removal of a hemorrhoid, probing of an aching ear, or a Caesarian delivery, medieval Muslim medical practice was governed by rigorous ethics regulating the relationship between physician and patient and controlling surgical and clinical procedure. This system of ethics was backed by arduous training that was more comprehensive and advanced than any carried on at the time in Western Europe.

MEDICINE

Figure 9.8
Physician Treats a Blind Man, MS Illustration from a Thirteenth-century Copy of Dioscorides' *De Materia Medica*, Iraq

This illustration of a physician treating a blind man is one of hundreds contained in a thirteenth-century Arabic translation of one of the most celebrated of all Greek medical texts, Dioscorides' *De Materia Medica*, written in the first century AD. In this treatise, comprised of five books, the author describes in full detail approximately five hundred plants that he studied while on military service with the Roman army in Asia Minor. Various versions of this work contain illustrations that deal with medical subjects other than medically useful flora. Dioscorides' texts were among the first to be translated into Arabic and had considerable influence on medieval and later medical practice, within Islam and beyond. The text in the chapter containing this painting deals with an herbal preparation for, among other things, headache. The physician's assistant, at the left, appears to be preparing for more aggressive treatment!

Figure 9.9
Physician and Attendant Preparing a Cataplasm (Poultice), MS
Illustration from Thirteenth-century Copy of *De Materia Medica*

This illustration, also from Dioscorides' *De Materia Medica*, shows a doctor instructing his assistant in the preparation of a poultice with a mortar and pestle. A similar pose of assistant and doctor (or student and teacher) appears in a number of medieval Muslim manuscripts, medical and other.

MEDICINE

Figure 9.10
A Medical Emergency: Diseased Dog Biting Man's Leg, MS Illustration from a Thirteenth-century Arabic Translation of *De Materia Medica*, Iraq

This illustration shows a diseased dog biting a man's leg. Many such life-threatening situations were observed and understood by Muslim physicians, who developed a great variety of antidotes and other curative procedures.

[151]

Figure 9.11
A Resourceful Victim of Snakebite, MS Illustration from *Kitab al-Tiryaq* (Book of Antidotes), Probably Iraq, Thirteenth Century

In this illustration from Galen's *Kitab al-Tiryaq*, a physician observes a boy, just bitten by a snake, undertaking to kill and eat the snake together with some handy laurel berries, thus effecting a cure. This celebrated book included medical writings collected from classical sources and translated into Arabic, most notably by Hunayn ibn Ishaq in the ninth century. The science of toxicology was significantly advanced in medieval Islam. Poisonings by enemies both human and reptilian were frequent enough to warrant the development of a wide variety of antidotes, many from natural or animal sources.

Figure 9.12
Dioscorides Handing Over the Fabulous Mandragora to One of His Disciples, MS Illustration from an Arabic Copy of *De Materia Medica*, Mosul, Iraq, Thirteenth Century

In this illustration, the Greco-Roman naturalist Dioscorides is shown giving a mandragora (mandrake) plant to one of his disciples. The European mandrake is found in Mediterranean lands and for thousands of years served as an anesthetic for surgery, or as a sleeping aid for a patient in pain. Much ancient and medieval superstition surrounded the plant, which was believed to have magical powers due to its resemblance to the human body. The mandrake has also been used to induce passion as well as conception, and to induce vomiting. Its modern application is limited to homeopathic prescriptions for such ailments as allergies and asthma, and it is recommended for use only under medical direction. Both the European and American versions of this plant are poisonous. This illustration, contained in a thirteenth-century Syrian or Iraqi manuscript of *De Materia Medica*, indicates that some medieval Muslim artists were influenced by pre-Islamic, Byzantine painting styles.

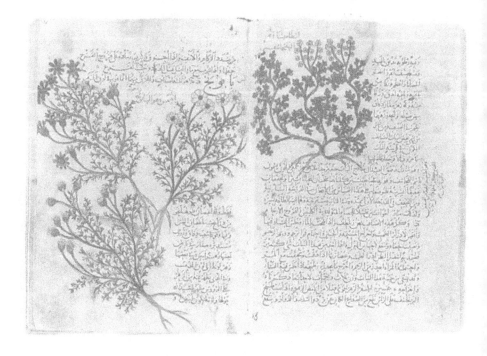

Figure 9.13
The Useful Chamomile, MS Illustration from *Al-Adwiya al-Mufrada* (On Simples) by Abu Ja'far al-Ghafiqi, Probably Spain, Thirteenth Century

This familiar herb has been used for centuries to deal with everything from upset stomach to a variety of nervous conditions, as well as neuralgia and rheumatism. This decorative page from a thirteenth-century copy of an Arabic manuscript by the twelfth-century Andalusian physician al-Ghafiqi, is one of more than three hundred and fifty colored renderings of plants and animals arranged alphabetically in an encyclopedic text entitled *On Simples* (medicinal plants). Al-Ghafiqi, a pioneer in medical botany, pharmacy, and materia medica, brought the identification of drug properties up to date.

Figure 9.14
Iris and White Lily, MS Illustration from an Arabic Translation
of *De Materia Medica*, Persia, Fifteenth Century

Arabic translations of *De Materia Medica* formed the basis for much innovation achieved by eminent Muslim scholars and physicians in pharmacology and pharmacy. This effort involved a chain of varied practitioners, including naturalists, botanists, collectors, manufacturers and sellers of medicinal herbs and spices, trained apothecaries, and experienced pharmacist-authors.

Figure 9.15
Medieval Muslim Pharmacy, MS Illustration (Detail) from an Arabic
Translation of *De Materia Medica*, Iraq, Thirteenth Century

This view, from a thirteenth-century Arabic manuscript of *De Materia Medica*, shows pharmacists preparing medicine from honey; above them an attendant appears to be checking a sample next to the storage area. The medieval Muslim pharmacy, whether as a separate neighborhood store or as a facility within a hospital complex, was far superior in resources and techniques to the equivalent in medieval Western Europe. Field-trained apothecaries were also an important link in the chain of practitioners involved in the operations of pharmacology and pharmacy. Muslim influence on later Western developments in these disciplines was considerable and is reflected in today's renewed interest in natural medicine.

MEDICINE

Figure 9.16
Pharmaceutical Packaging in Medieval Muslim Lands: A Drug Jar
(Alborello), Ceramic, Syria, Late Thirteenth Century

Glazed and underglaze-painted ceramic drug jars, or alborelli, such as this example, display a notable Muslim talent for endowing utilitarian containers with elegance and beauty. Developed to a high degree in the Near East, such skill spread first to Muslim Spain and then to late Medieval and Renaissance Italy.

Figure 9.17
Anatomical Study of the Horse, MS Illustration,
Egyptian, Fifteenth Century

This study was included in a fifteenth-century Egyptian manuscript probably devoted to veterinary practice. Arabs have a long and outstanding tradition of horse-breeding, and a number of influential texts were produced by specialists such as Ibn Akhi Hizam, a groom at the royal stables in Baghdad in the ninth century, and, in particular, by Abu Bakr al-Baytar, chief groom and veterinarian at the court of Al-Nasir Muhammad Ibn Qalaoun, a fourteenth-century ruler who considered the horse to be a sacred and honorable creature. Medieval Muslim veterinary medicine covered many aspects of caring for horses, including exercise and physiology, as well as deformities, diseases, and their treatment.

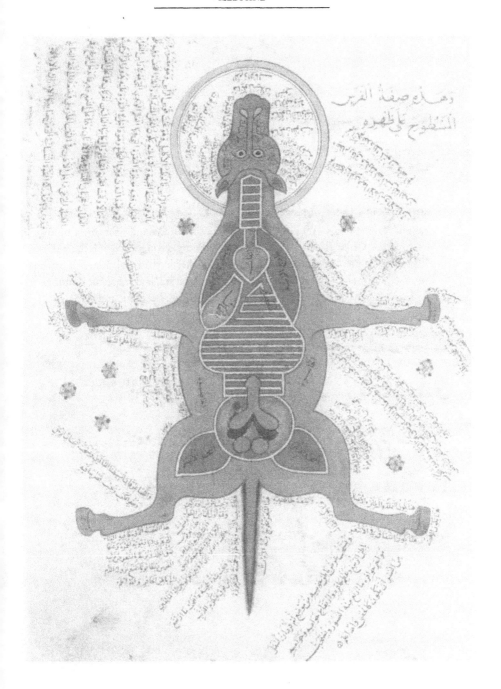

Figure 9.18
Bath Scene, MS Illustration from *Haft Awrang* (The Seven Thrones) by Jami, Persia, Sixteenth Century

This scene from a sixteenth-century Persian manuscript, *Haft Awrang*, by the classical poet Jami, reveals a traditional institution, the bath, in one of its most sumptuous modes. Quite apart from the required ablutions connected with daily religious practice, the Muslim communal bath, whether in the royal palace or town center, has always provided healthy and social relaxation, and it has for centuries been adapted to provide special therapy in Muslim hospitals and clinics.

MEDICINE

10

Natural Sciences

The pursuit of knowledge in the Islamic world has historically been shaped by a number of interacting, sometimes opposing, factors, religious precept, practical need, and the powerful influence of inherited and adapted cultures among them. In particular, the intellectual treasury of classical Greece and Rome, filtered through Hellenistic thought and practice, dominated much of the heritage that was appropriated and adapted by most medieval Islamic philosopher-scientists.

On the one hand, as has been noted, the idea of deliberately pursuing any kind of knowledge merely for its own sake, so familiar to the Western world, was considered alien, even impermissible, by virtually all of Islam's orthodox religious leadership and by many of its believers. Gaining knowledge was valid only, or at the very least principally, as a step toward the attainment of the just and holy life prescribed by the Prophet. On the other hand, the worldly, humanistic points of view represented in the works of ancient Greece's great thinkers, beginning with Aristotle and Plato, exercised a profound, often leavening influence on Muslim intellects.

As we have seen, mathematical and astronomical principles largely obtained from the "ancient sciences" of the pre-Islamic world, both West and East, were, once introduced to Muslims, quickly utilized in the procedures required for religious orientation and practice in a specific time and place. Very shortly these exact sciences were applied in solving the everyday physical problems of creating and extending the superstructure of the expanding Muslim empire—its roads and bridges, its forts, palaces, and cities, along with the welfare of its peoples. Perhaps most important, examination of the broad pursuits of the most eminent Muslim philosophical and scientific thinkers reveals clearly a range and intensity of focus that reflects more than spiritual motivation, more than utilitarian application. In short, from the beginning of Islamic civilization, devotion to knowledge for its own sake, while anathema to the *ulema*, the orthodox Muslim religious leaders, was a crucial factor in much Muslim intellectual and scientific achievement.

What evolved after the tenth century was a change in the character of the Islamic scientific effort: the boundaries between "ancient" and "religious" sciences, between Hellenistic and orthodox Islamic approaches to investigation and speculation, became less distinct and less contested. Meanwhile, one fundamental precept applied from the start, whether framed in Hellenistic or Islamic mode: Each person's rightful task was to put together the pieces of a universal puzzle, mindful that there was a place for everything and that everything was connected. Thus, just as much attention had to be paid to earth as to people, whose physical welfare depended on what they could gain from the soil and from the waters. Muslims seem to have grasped the importance and the complexity of careful use of the earth's resources long before Europeans did. Perhaps the rapid Muslim expansion across the vast territories between the Atlantic and Indian oceans and the development of these lands made good environmental behavior especially relevant to survival.

As their domain expanded east and west, Muslim naturalists eagerly investigated the rocks and soils as well as the flora and fauna of every region from Spain to western India and beyond, compiling a lengthy inventory that was without equal in the medieval West. Detailed reports and analyses were produced in abundance: *On the Horse, On the Camel, On Wild Animals, On the Vine and Palm Tree, On the Making of Man*. Nothing zoological or botanical escaped notice, analysis, and thorough classification. The character of much of this work was influenced by Aristotelian concepts inherited from ancient Greece, with each class of items hierarchically ordered and fixed. Ibn Sina, al-Biruni, and al-Khazini, a former Greek slave who lived in Persia in the twelfth century, measured and classified precious and semiprecious stones. Ibn Sina also investigated geology and the influence of earthquakes and weather. The tenth-century geographer-encyclopedist al-Masʿudi (celebrated also as a philosopher and natural historian) even provided the beginnings of what might be called a theory of evolution.

Much of the botanical investigation carried out by Muslims directly benefited the pharmacology and pharmacy that developed throughout the Islamic world on an unprecedented scale. This harvest of facts concerning all kinds of animate and inanimate species prompted the production of a variety of encyclopedias. These were often profusely illustrated, and some of the most colorful ones displayed not only what naturalists and other scientists had actually examined but also what they had simply learned at second hand, sometimes from sources inspired by fancy. The celebrated thirteenth-century work of Zakariya ibn Muhammad ibn Mahmud Abu Yahya al-Qazwini, *Ajaʾib al-Makhluqat* (The Wonders of Creation), reveals some more or less realistic cosmology, botany, and zoology along with splendidly imaginary items, such as trees that sprouted birds rather than leaves. Since this phenomenon was claimed to exist only in the British Isles, virtually none of al-Qazwini's readers could attempt to verify it in person. Of course, most Muslim ornithologists focused on the actual inhabitants of the skies, and there are a number of medieval Arab works on the princely art and sport of falconry.

The Muslim store of knowledge concerning flora and fauna in the lands they controlled, together with experience in cultivating a wide variety of foodstuffs, enabled agriculture to thrive, especially in Western areas such as Spain, which was, in this as in other ways involving the sciences, far in advance of the rest of Europe. In Andalusia and elsewhere the Arabs replaced the ox with the more efficient horse. Moreover, the extensive network of trade routes established throughout Islam promoted the distribution of many items far from their places of origin. Europeans knew only honey as a sweetener until Muslims introduced sugar to the West. Other exports introduced from and via Islamic lands included oranges, rice, and many spices, as well as the large number of medicinal plants previously mentioned.

Although a broad survey of Muslim technology, including Muslim innovations in craft and construction techniques, both civil and military, is beyond the scope of this book, it seems appropriate here to give some attention to ways in which Muslims developed solutions to problems related to the one natural element that concerned them most, especially in connection with agriculture: water.

The never-ending preoccupation with water in the Islamic world is understandable. The earliest Muslim homeland, the Arabian peninsula, consists mostly of desert and parched highland. Muslim territories in the Near East and North Africa are almost as dry. Even the large regions at the far reaches of the Muslim conquest—Spain and Western India—have never been without excessively dry seasons, nor are they unfamiliar with drought.

Providing people with water, Muhammad is believed to have said, is the act of greatest value. Guided by the findings of Archimedes and other hydraulic specialists that preceded them, Arabs made many improvements in the waterwheel, learned to construct wells with fixed water levels, and built elaborate irrigation networks, especially in Mesopotamia and Egypt. In Persia, Muslims also increased the scale and capabilities of the *qanat*, a type of underground conduit that links together a series of wells and is capable of tapping distant sources of ground water.

Gaining ingenuity in finding water and in utilizing it for survival, Muslims eventually found ways of exploiting it for smaller-scale purposes. Arabs produced sophisticated versions of the water-clock, a device that dates back to antiquity (see Figs. 10.12a, 10.12b, and 10.13). A broad range of mechanical devices designed by the celebrated ninth-century scientists the Banu Musa included apparatuses for providing hot and cold water, well-construction aids, and a self-trimming lamp. Descriptions of such items were precisely spelled out in one of the Banu Musa texts, *Kitab al-Hiyal* (On Mechanical Devices), which, translated into Latin, helped to transmit Muslim technological expertise along with ancient Archimedean concepts to the West.

In addition, a great deal of attention was paid to improving methods of observation and analysis as well as balance and measurement techniques. The versatile al-Biruni devised a number of advances in methods of determining specific gravities. Al-Khazini produced a long treatise, *Kitab Mizan al-Hikma* (Balance of Wisdom), that not only records specific gravities of solids and liquids, but also establishes standards of measurements, discusses balances, and sets forth theories of capillarity and ingenious systems of leverage. This text became a standard reference in medieval Europe.

Among the many original Muslim mechanical devices known as "automata," perhaps the most unusual and ingenious hydraulic examples were designed by Badiʿ al-Zaman al-Jazari, a twelfth- to thirteenth-century engineer who appears to have been employed in the service of a series of rulers of the Artuqid dynasty in southeastern Turkey. Most of his large-scale mechanisms were designed to collect and transport water; among them are complex pumps which involved scoops, wheels, gear systems, transport jars, and ox-power. This master craftsman is also celebrated for his exotic small-scale devices which kept time, measured blood, or poured wine and other liquids, and, whether the function were serious or frivolous, generally astonished and delighted onlookers.

NATURAL SCIENCES

Figure 10.1
A Man Gathering Plants, Drawing (Detail), Persia, Seventeenth Century

This seventeenth-century drawing from Persia presents an activity that was widespread in Muslim lands from the beginning of Islam. Prompted both by a fundamental concept that understanding God requires learning about all living and inanimate things, animal, vegetable, and mineral, and by the everyday needs of survival and progress, often within environments far from hospitable, Muslim naturalists, physicians, botanists, and zoologists accomplished within a few centuries an encyclopedic investigation into the nature of the world around them.

Figures 10.2 and 10.3
The Fauna of Muslim Lands

These fifteenth- and seventeenth-century illustrations, from Persia and Turkey, demonstrate some of the differences in artistic style that have characterized Muslim pictorial records of the natural world. The detailed representation revealed in these examples suggests the precise naturalistic approach that can be found in the works, produced two centuries later, of artist-ornithologist John James Audubon.

Figure 10.2
Small Black and White Bird on a Limb with Butterflies

Detail of mid-seventeenth-century (?) painting by Shafi ʿAbassi, Iranian, Isfahan; Color on paper, 16.2 × 24.9 cm.

NATURAL SCIENCES

Figure 10.3
Hunting Hawk, MS Illustration from *Album of Mehmet II*,
Turkey, Fifteenth Century

Figure 10.4
The Natural Encyclopedists: *Leopard*, MS Illustration from *Manafi˓ al-Hayawan* (On the Identification and Properties of Animal Organs) compiled by Abu Sa'id Ubaydallah ibn Bukhtishu˓, Persia, Thirteenth Century

A considerable production of illustrated encyclopedias by medieval Muslim scholars greatly expanded Islam's inheritance of ancient botanical and zoological learning. On the opposite page is a folio from a thirteenth-century copy of a tenth-century text, *Manafi˓ al-Hayawan*, compiled by ibn Bukhtishu˓, a Christian who was court physician at Baghdad. This manuscript deals with a great variety of animals, both real and imagined, describing their characteristics and qualities, as well as the therapeutic uses to which their various organs might be put. Combining useful fact with charming (and sometimes startling) fancy, the work occasionally reads like an old-fashioned home remedy almanac, at other times like a fairy tale. For example, the author states that leopard's meat and fat, boiled in olive oil, make a good ointment for skin blemishes and outbreaks; he also claims that carrying a wolf's right eye will serve as protection against the evil eye and other magic spells.

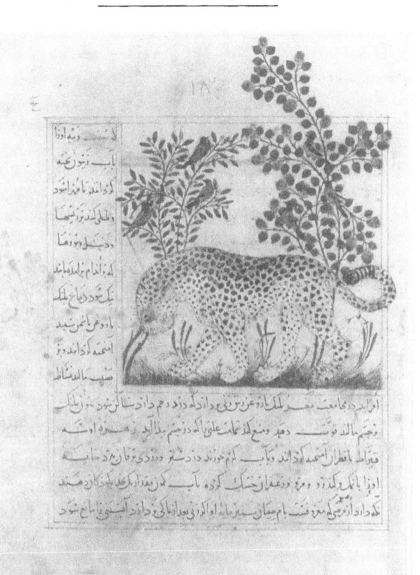

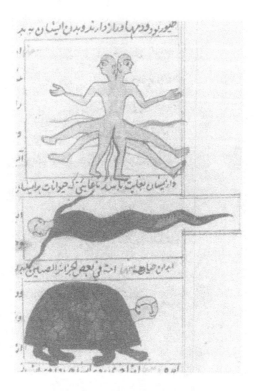

Figure 10.5
The Natural Encyclopedists: Selection of Fanciful and Realistic Fauna and People or Creatures, MS Illustration from *Aja'ib al-Makhluqat* (The Wonders of Creation) by Zakariya ibn Muhammad ibn Mahmud Abu Yahya al-Qazwini, Iraq, Fourteenth Century

One of the most extraordinary of all medieval Muslim encyclopedias, *Aja'ib al-Makhluqat* was completed in Iraq in the thirteenth century by al-Qazwini, a cosmologist and geographer whose two-volume work covers virtually everything on earth and beyond it as well. Heavenly bodies, including angels, the earthly elements, and all living organisms, not to mention geographical details of many lands and seas, are discussed and profusely illustrated. Here, fact and myth are well mixed and presented with conviction, if not always with proof. In the book can be found, for example, a description of birds that grow on trees, along with fearsome details of other phenomena, such as a Tibetan animal a glimpse of which brings death to the observer!

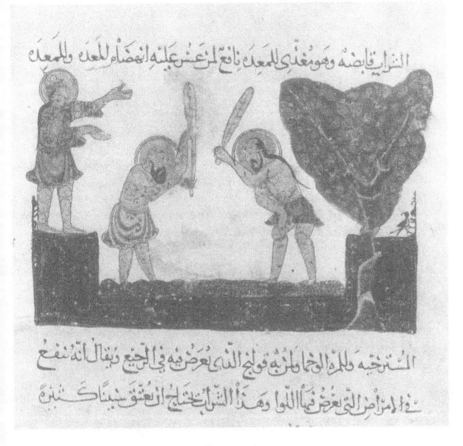

Figure 10.6
Men Treading and Thrashing Grapes, MS Illustration from *De Materia Medica*, Arabic Translation, Iraq, Baghdad School, Thirteenth Century

The widespread investigation of soils in all the Islamic lands, as well as the study of the minerals, conditions of climate, seasonal changes, and ecological differences between one area and another, promoted a remarkably advanced horticulture and agriculture. The resulting knowledge, transmitted to Europe after the eleventh century, helped to improve farming techniques, widen the variety of crops, and increase yields on the continent's farmlands. In addition, an enormous variety of crops was introduced to the West from or through Muslim lands, ranging from sugar, oranges, and melons to coffee, sesame, and cotton.

Figures 10.7a, 10.7b, and 10.7c
The Technology of Water

For centuries, the dry and harsh ecology of much of the Muslim lands has made the collection, transportation, and storing of water a priority. It is hardly surprising that the most important progress in medieval Muslim technology and engineering was achieved in relation to water. Waterwheel mechanisms, invented centuries before Islamic civilization began, were improved by Muslims, who also developed resourceful methods of water utilization, including the manufacture and storage of ice.

The reservoir shown in Figure 10.7b was built in Kairouan, Tunisia, in the ninth century. Figure 10.7c shows the seventeenth-century bridge and dam built in Isfahan to control the waters of the Zayandeh River. Muslims also developed extensive irrigation systems of underground channels, or *qanat*, which could transport subterranean water sometimes a dozen miles or more from its source.

Figure 10.7a
Waterwheel in Action, Hama, Syria

Figure 10.7b
Ninth-century Reservoir, Kairouan, Tunisia

Figure 10.7c
Khvaju Bridge, Isfahan, Iran, Showing Sluice Gates, Seventeenth Century

Figures 10.8 and 10.9
Water——on Earth and in Paradise

Through the centuries, the efforts of finding, collecting, and transporting water have been of the greatest importance in most Muslim regions, not only for material survival and progress in largely arid land; they have also filled profound emotional as well as spiritual needs. In millions of homes throughout the Islamic heartland, returning from the tasks of the day has often meant returning to a home centered around a courtyard, itself centered on a small fountain, the sight and sound of which has provided respite from the burdens of work and a harsh environment. On a larger and more public scale, water has also made possible the extraordinary development of Islam's monumental gardens, which, especially in Andalusia and India, have created an almost unearthly world of enchantment, one that sometimes appears intended to create, as best as human beings could achieve it, a model of what Paradise should be, a vision of heaven on earth.

Figure 10.8
Farmers and Animals, MS Illustration from *Kitab al-Tiryaq* (Book of Antidotes) by Pseudo-Galen, Northern Iraq, Twelfth Century

Figure 10.9
Traditional Muslim Gardens and Fountains, Generalife, Alhambra,
Granada, Spain, ca. Fourteenth Century

Figure 10.10
A Muslim Tradition Still Alive in Valencia:
Present-day Valencia Water Court Meeting

In the tenth century Arabs developed an irrigation system in Valencia, in Spain, that still functions today. A group of citizens known as the Water Court, the *Tribunal de las Aguas*, meets at noon every Thursday in front of the main cathedral to assess and adjust as necessary the current allocation of the region's water supply to the city's neighborhoods. The weekly ceremony, also established under Arab rule, involves the hearing of disputes and the handing down of decisions in response to which there is no appeal. Islamic Andalusia, along with Sicily, served as an important gateway through which agronomic as well as many other Muslim scientific developments were passed to medieval and Renaissance Europe.

NATURAL SCIENCES

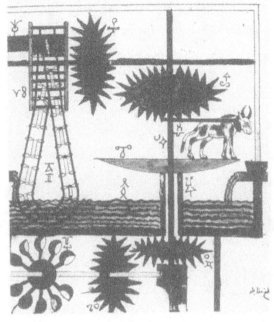

Figure 10.11
Al-Jazari's Design for a Water-raising Device, MS Illustration from
Al-Kitab fi Ma'rifat al-Hiyal al-Handasiyya (The Book of Knowledge
of Ingenious Mechanical Devices) by Badi' al-Zaman al-Jazari,
Fourteenth Century

This pump, designed by al-Jazari, the greatest medieval master of the mechanical arts, served as a model for an actual lakeside machine that was used in the thirteenth century in Damascus. In the actual device, water from a lake turned a scoop wheel and a system of gears that transported jars which transferred water up to a channel that led to a hospital and mosque. In the model, which was used as a decorative amusement, the gears were hidden, and the device appeared to be driven by an ox rather than by water-power. Al-Jazari's considerable output included a great variety of utilitarian hydraulic and other devices including clocks, automated gates, locks and bolts, and other mechanisms. He also produced many devices known as "automata," possessing considerable elegance and charm and intended mainly for simple enjoyment.

Figures 10.12a and 10.12b
Traditional Outdoor Water Clock, Opposite Bu'Inaniyyah Madrasa, Fez, Morocco.

Known to have existed in ancient Babylon, clocks operated by waterpower were developed further in Greece, China, and India. These photographs show what remains of a very large medieval example under restoration in Fez, Morocco.

Figure 10.13
Design for a Castle Water Clock, MS Illustration from
Al-Kitab fi Ma'rifat al-Hiyal al-Handasiyya (The Book of Knowledge
of Ingenious Mechanical Devices) by Badi' al-Zaman al-Jazari,
Mesopotamia, Fourteenth Century

Inheriting a sophisticated tradition of mechanical design, the ingenious al-Jazari took it to new heights of inventiveness in the twelfth century, as is indicated by this illustration of his text on mechanical devices. Some of his clocks, monumental in scale, displayed moving models of the sun, moon, and stars, as well as a virtual circus of performing musicians and other figures, human and animal. The clock shown here, when built, stood about eleven feet tall, featured displays of the zodiac, as well as of the solar and lunar orbits. A pointer traveled across the top of the gateway and caused doors to open in sequence on the hour, which was sounded by a pellet dropped from each of two falcons, striking cymbals within bowls beneath them.

Figure 10.14
Basin of the Two Scribes, MS Illustration from *Automata*,
by Badiʿ al-Zaman al-Jazari, Syrian Copy, Fourteenth Century

This illustration shows al-Jazari's design for a medical device to be used in measuring the amount of a patient's blood taken during bloodletting. Blood running into a basin increases its weight, thus operating a pulley which causes the two scribes above to rotate; one scribe indicates a marked point on a calibrated circle, the other points to a tablet marked in the same way.

Figure 10.15
Design for Water Fountain of the Peacocks, MS Illustration from
Al-Kitab fi Ma'rifat al-Hiyal al-Handasiyya (The Book of Knowledge
of Ingenious Mechanical Devices) by Badi' al-Zaman al-Jazari, Copy,
Mesopotamia, Fourteenth Century

This al-Jazari device may well have intrigued, as well as served, elegant medieval Muslim diners. A guest was to hold out his hands in front of the peacock, which delivered a little water onto them. A mechanical slave would emerge from the castle, offering some powdered soap. When the water stopped running, another slave would appear with a towel.

Figure 10.16
Mechanical Boat with Drinking Men and Musicians, MS Illustration from *Al-Kitab fi Ma'rifat al-Hiyal al-Handasiyya* (Book of Knowledge of Ingenious Mechanical Devices) by Badi' al-Zaman al-Jazari, Copy, Iraq, Thirteenth Century

One of a series of al-Jazari "automata" designed to amuse the guests at a drinking party, this rather urbane and courtly toy, when activated, came alive, with sailors rowing and musicians playing intermittently. Much of the importance of al-Jazari's technical accomplishments, covering many kinds of mechanisms, many of them seriously useful, consists in the thorough and craftsmanlike way he described the purpose of each device's component parts, the way in which each part was to be made and assembled with the others, and the way in which the device was to be tested. His work still speaks eloquently today, especially to anyone who likes to invent things, let alone tinker with them.

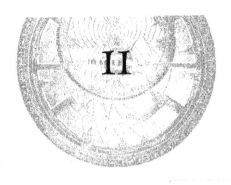

II

Alchemy

Centuries before the rise of Islamic civilization the fundamental premises of alchemy were fairly well established: all substances are composed of the four elements mixed in various proportions; gold is the most noble and pure of all metals, followed by silver; it is possible to transform one metal into another by changing the mixture of elements that it contains; a base metal can be transformed into a noble one through the use of the fifth element, called the elixir. Alchemists also believed that inorganic substances were alive, composed of spirit as well as matter. The combining of substances involved the union of the male and female principles of sulphur and mercury (an echo of the Chinese yang and yin complementaries). The two components could be separated by heat; the resulting vapor, the spirit, could sometimes be condensed into liquid, which contained the essential qualities of its source. This liquid, when transferred to a less noble material, gave the lesser material nobility. Underlying these beliefs and deeply embedded in medieval alchemic philosophy was concern with the spiritual development and salvation of humankind.

Alchemy combines spiritual, craft, and scientific disciplines that can be traced back to very ancient times and to processes traditionally involved in early metalworking and drug preparation. Chinese, Indian, and Greek philosophical and metaphysical concepts, originally focused on the fundamental nature and potential of metals and earth compounds, began to guide the application of crafts in the pre-Christian era beyond practical ends such as metal implements and drugs. By the time of the Hellenists there was a distinct difference between alchemy and rudimentary chemistry, even though the experimental apparatus and basic experimentation involved in these sciences were fairly similar. Medieval interest in magic and in mysticism, the belief in a reality beyond that perceptible by the ordinary senses or intellect, set alchemy near the forefront of the intellectual arena. Practical chemistry's time in the sun would begin only after alchemy had lost some of its spiritual strength and allure, sometime around the end of the fourteenth century.

Once the civilization of Islam was extensively established, Muslims absorbed the basic alchemical canon as shaped by the Alexandrians and proceeded to reshape it within their own intellectual conventions. Alchemy came to be regarded as an occult science that dealt with properties of matter whose causes could not be perceived by the senses. Two specific definitions evolved: alchemy was "the art of reproducing nature," as well as the "science of the balance." Muslim alchemists believed that precious metals such as gold could be made by observing, improving, and applying the methods of nature. The properties of an element (fire, air, earth, or water) that existed in a given mineral could be determined through numerology, the study of the occult meanings and influences of numbers. These properties could then be brought into balance through alchemy. As for the long-sought end product—the elixir, or "philosopher's stone," it would be better than gold because it could transmute other metals into gold. Thus did Muslims adapt inherited alchemical notions in ways that reflected both practical and mystical aims.

Among Islamic explorers of the labyrinthine passages of alchemy were a number of distinguished scientists, including members of the Ikhwan-al-Safa', the Brethren of Sincerity. This secret tenth-century mystic sect was composed of philosophers with a remarkably tolerant view of religions other than Islam. They wrote an encyclopedia that embraced natural science topics, such as mineral formation, earthquakes, and meteorological matters, all connected with the heavenly bodies and the celestial sphere. The Brethren were opposed by the religious authorities, and the group went underground in the eleventh century. Its radical and relatively rational theories, however, including Neoplatonic and universalist ideas and an allegorical interpretation of the Qur'an, spread as far as Spain and Central Asia, where they influenced philosophic and scientific thought. Echoes of the Brethren's teachings, somewhat beyond the mainstream of Islamic thought, survive today, mostly among esoteric spiritual groups in Asia Minor and India.

An enormous body of writings has been attributed to the possibly legendary Jabir ibn Hayyan, known as Geber in the West, an eighth-century Baghdad alchemist. Among his central concerns were said to be the principle of the "balance" and its system of numerical relationships between elements in substances, as well as the four elemental natures of things: heat and cold, dryness and humidity. He apparently specialized in types of elixirs and their various capabilities of transmutation. He seems to have been not only a superior and practical laboratory master, one who provided clear directions for lab procedures, but also a thorough observer and analyzer. He was an innovator as well: he was said to know about ways of producing steel, dyeing leather and fabrics, and waterproofing clothing. No matter that the actual source or sources of the Jabirian corpus are debated, the work embraces much of what was known of alchemy at the time the treatises were written.

The great ninth-century physician al-Razi involved himself deeply in alchemy and, perhaps more than any other practitioner in this field, demonstrated a firm preference for proof through experiment, as opposed to wholly theoretical or magical procedures. In his hands, basic alchemical processes such as distillation, calcination, crystallization, evaporation, and filtration gained precision. The array of laboratory utensils and vessels he employed was expanded and refined until the standard alembics, beakers, flasks, funnels, and furnaces began to resemble those of modern times. Al-Razi surpassed Jabir in systematically classifying substances, such as minerals, and in including in his category of substances those that were obtained artificially in the laboratory along with those he classified as natural. His alchemy appears to have veered ultimately toward chemistry. In the fourteenth century, the historian, sociologist, and philosopher Ibn Khaldun took a long look at alchemy and delivered a firmly negative verdict, demonstrating that alchemy is harmful to people and that its efficacy has never been proved.

The science of chemistry as practiced in the West after medieval times owes more to Muslim alchemy than tools, techniques, and a lexicon of alchemical terms such as *al-iksir* (which became "elixir"). The beginnings of the experimental method may be traced in part to the ways in which Jabir, al-Razi, and other medieval alchemists approached and carried out their work. However, Muslim alchemy should not be thought of as a primitive and spiritual version of what later developed into the secular and empirical disciplines of modern chemistry, which, even as it endeavors to improve living conditions, has nothing to do with probing the "souls" of materials or with saving those of human beings. The assumptions and goals of medieval alchemy were founded in sacred and semi-sacred belief, ritual, and magic. For all but serious mystics, the science of chemistry is founded on curiosity, practical necessity, and objective, trial-and-error experiment. Most medieval alchemists served both alchemical and chemical purposes without being conscious of their fundamentally disparate characters. Their scientific achievements lay largely in the advances they made in experimental techniques and procedures. It was these advances that played a part in setting the stage for the chemistry of the modern age.

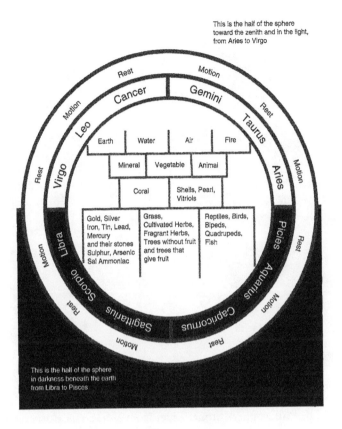

Figure 11.1
Diagram of the Cosmology of Alchemy

Medieval Islamic alchemists inherited from ancient civilizations many basic concepts about matter and how it could be transformed, and they adapted them to fit within a Muslim mystical or metaphysical philosophical framework. Alchemy became defined as an occult science that dealt with properties of matter whose causes were imperceptible by the senses. It was believed that alchemists could determine the proportion of elements in minerals by numerology and, through laboratory procedures, bring them into superior balance, even transmuting them into gold. This diagram, reflecting the basic tenets of celebrated alchemists such as the possibly legendary Jabir ibn Hayyan (who has been described as an eighth-century non-Muslim pagan from Syria), places the elements and all things inanimate and animate in a dynamic, zodiacal framework. Jabir has been said to have been primarily concerned with probing the "souls" of materials, as well as with saving those of human beings.

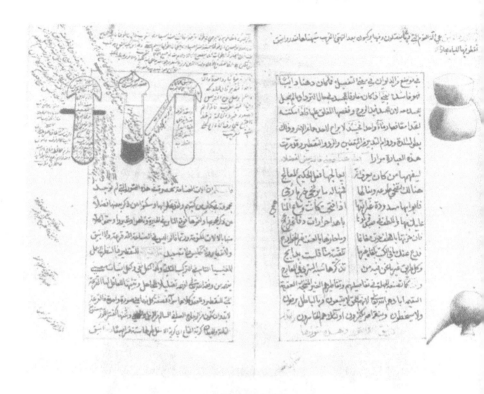

Figure 11.2
The Philosopher's Stone: Pages from *Sharh diwan al-shudhur* (Commentary on a Collection of Poems about the Philosopher's Stone) by 'Ali ibn Musa ibn Arfa-Ra's, Twelfth Century

The manuscript shown here contains poems about the philosopher's stone: *al-iksir*, the elixir, alchemy's long-sought end product, better than gold because believed capable of transmuting other metals into gold. The diagrams and illustrations of alchemical apparatus shown here indicate the typical laboratory procedures involved in medieval alchemy. In the hands of Jabir, al-Razi, and others, such processes as distillation, evaporation, and filtration became greatly refined, and furnaces, alembics, flasks, and other implements began to resemble those used in modern labs. While medieval alchemy should not be regarded as a primitive, superstitious version of what later became the secular and entirely empirical science of chemistry, the beginnings of today's experimental methodology may be traced in part to ways in which medieval alchemists approached and carried out their tasks.

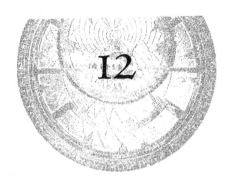

12

Optics

Several of the most eminent medieval Muslim philosophers, mathematicians, and physicians devoted extensive study to the fundamental nature and workings of vision and of light. It was in the field of optics that they made what are probably the most original and important scientific findings in the history of the Islamic world.

These scientists had access to a rich treasury of Greek knowledge concerning light and vision, including, most notably, works written by the mathematician Euclid in the third century BC and treatises that the Egyptian astronomer Ptolemy produced four hundred years later. This pre-Islamic literature explored an encyclopedic range of topics, from reflection, refraction, image projection through openings, and the rainbow, to the anatomy and workings of the eye. Greek treatises in these fields had utilized the terms of several disciplines, including mathematics, natural philosophy and medicine. It was some time before Islamic scientists, building from this base, developed a more unified approach to optical phenomena as a whole.

Working during the ninth century with optical theories obtained from Euclid's *Optics*, al-Kindi produced a new understanding of the reflection of light, as well as of the principles of visual perception, the beginnings of what became, in the European Renaissance, the laws of perspective. Determined to reconcile the elements of natural science and mathematics, al-Kindi rejected the Aristotelian concept of vision as form received by the eye from the object being looked at. Instead, he perceived vision as generated by a luminous force that travels from the eye to the object in the form of a solid cone of radiation.

Two later philosophers, al-Razi and Ibn Sina, also dealt in their texts with optical matters. However, the high quality of the Muslims' systematic research in vision and light is demonstrated most spectacularly by the achievements of Ibn al-Haytham, born in Iraq in the tenth century and known in the West as Alhazen. Of all known medieval scientific texts, his comprehensive *Kitab al-Manazir* (Book of Optics) is perhaps the most distinguished example in terms of its experimental and mathematical arguments, as well as in its presentation of new and original theory. Ibn al-Haytham investigated virtually every aspect of light and human sight. He studied the way light is refracted, or bent, by water, air, and mirrors. He came close to a theory of magnifying lenses. He examined the rainbow, aerial perspective, and sunlight. He explained correctly why the sun's and moon's diameters appear to increase as these bodies approach the horizon (an optical illusion caused by their expected size in relation to familiar objects on the ground). He also demonstrated how refraction by the atmosphere causes the sun to be still visible when it is actually below the horizon.

In order to study the solar eclipse, Ibn al-Haytham cut a small hole in a wall, allowing the semi-obscured solar image to be projected through it onto a flat surface. This early example of "camera obscura" optics anticipated modern photographic principles, just as his focusing experiments using parabolically shaped burning mirrors pointed the way to the lenses of future telescopes and microscopes.

Investigating the human eye, Ibn al-Haytham studied its structure, analyzed stereo vision, and formulated the method by which we receive images. Theories of vision held by such ancients as Euclid, Ptolemy, and Aristotle had offered various explanations for sight. One theory held that the eyes emit rays of fire which meet some sort of emanation from the object being looked at; together these emissions produce an image of the object. Another suggested that the eyes emit rays which hit the object and make it visible (not so very far from radar!). Yet another proposed that objects send out rays in all directions; some come through the pupils of the eyes and bring about an image of the objects. Ibn al-Haytham agreed in good measure with this last view but enriched it in original ways: he regarded vision as occurring when rays from the perceived object, composing a "form" representing an object's visible features, enter through the pupils (which act as lenses) and proceed to the brain, where the faculty of sense completes the process.

In Ibn al-Haytham's model of vision, the eye was involved as an optical system, in which psychology played an important part. This was an important advance beyond the theories of al-Kindi, who had supported many of the views of Euclid and Ptolemy, and it is not far from what we believe today. Ultimately, Ibn al-Haytham's conclusions exerted a strong influence on the work of da Vinci, Kepler, Roger Bacon, and European scientists in general. Among the most notable and original achievements of his *Kitab al-Manazir* is the fact that it demonstrates the integration of Euclidean optics with the perception of form found in Aristotelian physics. Here was a new and firm basis for subsequent investigation of vision: optics had now become a mathematical discipline. This "mathematization" of physics was a significant step in the genesis of modern science.

Muslims continued to make advances in understanding visual phenomena long after the eleventh century. Little attention, however, appears to have been paid to Ibn al-Haytham's *Kitab al-Manazir* until the thirteenth century. At that time, his formulation of the principles involved in the effects of the camera obscura was the object of an important commentary written by the Persian physicist Kamal al-Din al-Farisi (see Fig. 12.2). More important, al-Farisi provided the first satisfactory explanation of the rainbow, a phenomenon which had already fascinated Muslim scientists, such as the Brethren of Sincerity. Studying the path of rays of light inside a glass sphere revealed to him how sunlight was refracted through raindrops and thus how primary and secondary rainbows were formed.

The work of Ibn al-Haytham and al-Farisi greatly advanced the development of method in experiment, in particular the importance of correlation between experiment and theory. Slowly but inevitably, experimental science was taking shape, utilizing the kind of investigative process that ultimately came to dominate all scientific enterprise.

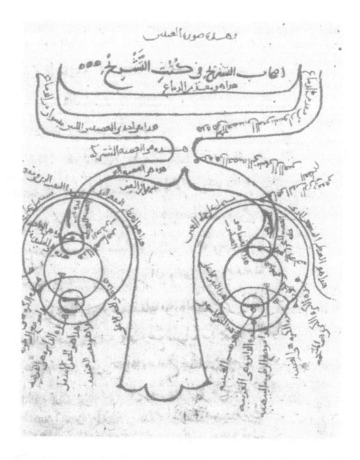

Figure 12.1a
Diagram of the Eyes and Related Nerves, MS Illustration from
Kitab al-Manazir (Book of Optics) by Ibn al-Haytham,
Istanbul, Eleventh Century

Of all the known medieval scientific texts, Ibn al-Haytham's eleventh-century *Kitab al-Manazir*, which includes this diagram, is perhaps the outstanding example of experimental and mathematical argument in the presentation of new and original theory. Building on the theories of Euclid, Ptolemy, and Aristotle, Ibn al-Haytham perceived vision as occurring when a "form" capable of representing an object's visible features enters through the pupil and proceeds to the brain, where the faculty of sense completes the process. This proposition is not far from what we understand of the process of vision today.

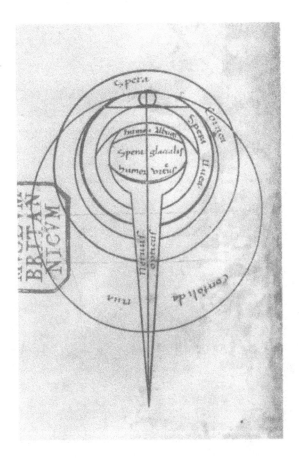

Figure 12.1b
Diagram Representing Ibn al-Haytham's Theory of Vision,
MS Illustration from a Fourteenth-century Latin Version of His
Eleventh-century *Kitab al-Manazir* (Book of Optics)

Ibn al-Haytham's concept of vision (displayed in Fig. 12.1a) is shown here in a more geometrically precise fourteenth-century illustration of his original text.

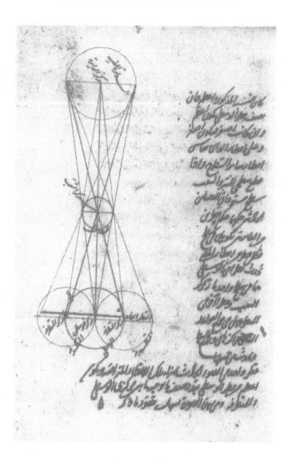

Figure 12.2
Diagram Illustrating Principles of the Camera Obscura, MS Illustration from a Résumé of Optics by Kamal al-Din al-Farisi, Istanbul, Fourteenth Century

Three centuries after Ibn al-Haytham's pioneering investigation of optics, the Persian al-Farisi extended Ibn al-Haytham's formulation of the principles underlying the phenomena of the camera obscura, familiar prototype of all photographic devices. Al-Farisi demonstrated that as apertures get smaller the images they focus get sharper; he also showed that inside the device the objects' image turns top to bottom and left to right. It was in probing the fundamental nature and workings of vision and of light that Islam's scientists made what are probably their most original and important findings.

13

The Later Years

By the beginning of the eleventh century, the civilization of Islam had reached the peak of its first golden age. The early Arab conquests had long been consolidated, despite dynastic upheavals and shifts. Muslims dominated the commerce of the Mediterranean basin. Great cities such as Córdoba in the west and Baghdad in the east attracted not only thousands of Islam's faithful bent on business and education but also increasing numbers of Europeans attracted by reports of opportunities, whether professional, educational, or financial, that were far more enticing than those in Christian Europe.

Córdoba, in particular, was an irresistible magnet, attracting many young, well-born Europeans, whose families sent them to the fabled Spanish metropolis to get "finished," much as American families eight centuries later would send daughters and sons to Western Europe's cultural centers. Eleventh-century Córdoba was a beautiful city with more than half a million inhabitants and three hundred public baths. One could move about safely at all hours through well-cleaned, stone-paved streets. Communal and private water supplies were reliable, the medical services beyond anything known north of the Pyrenees. Living standards, commercial opportunities, and cultural facilities here and in capitals such as Baghdad, Cairo, and Damascus were matched in quality, if not scale, by those in other

Islamic cities. Granada in Spain, Fez and Kairouan in North Africa, Palermo in Sicily, the holy cities of Mecca and Medina in Arabia, Ghazna near the Hindu Kush in northwestern India—these were only a few of the Muslim centers that shared the wealth of the thriving dynasties.

In such a realm, culturally diverse yet spiritually unified, ideas, innovations, customs, and trends spread quickly from region to region, community to community. Muslim arts had long flourished in virtually every region: the great early Arab mosques and the masterworks of craftsmanship in glass, plaster, and ceramics created in North Africa and the Near East were now followed by equally notable, if stylistically different, works by Persian, Turkish, and Indian architects and artisans. This artistic effort would continue triumphantly as late as the eighteenth century.

There was a moment in Islam's history, in the middle of the eleventh century, when the social and cultural achievements of the three previous centuries seemed relatively unthreatened by catastrophic political or social forces. To be sure, the next century would bring upheaval in the form of conquest by Seljuq Turks from Central Asia. And there would be new military advances in the west: Normans would take Sicily away from the Arabs, and the Christians would commence two hundred years of crusading in the holy lands against the infidel Saracens or Moors (as they called the Muslim enemy). The Christians would also, in the eleventh century, begin the reconquest of Muslim Spain. Islam as Empire would face disintegration.

Significant cultural change, however, let alone decline, was not yet apparent. Just as Muslim arts remained unsurpassed for several centuries, science was still vigorously pursued in certain disciplines, most notably astronomy, and in certain regions of the Islamic world. A considerable number of Muslim geographers, astronomers, mathematicians, physicians, and philosophers made important contributions as late as the fifteenth or sixteenth century. In sum, the late medieval years witnessed periods of enviable cultural stability or equilibrium in many Muslim regions. Yet seeds of both change and decline, some planted long before in Islam's early years, were taking root, and their ultimate yield would include strangling weeds.

A reminder about the general character of science in medieval Islam may be useful. Doubly impelled by faith and curiosity, the vigor with which early Muslim scientists went about their work is not surprising. The civilization that bred them was, during its first five centuries, more advanced than any other in existence at the time. Muslim scientists revered and welcomed the knowledge that they had extracted from the classical world, and they had learned to utilize in their own ways the intellectual and practical benefits of this largely Hellenic and Hellenistic legacy. There had long been developing a profound concern with the precise classification of knowledge, and this promoted a permanent and intense intellectual and theological debate.

The eventual categorization of science into two sets, one Islamic, the other "ancient" or "foreign," eventually led to a deep division in Muslim thought. The orthodox Islamic sciences, dealing with religion and related matters, and the ancient Greek sciences, devoted to pursuing knowledge of the physical world in a relatively rational manner, gradually became more clearly and actively perceived as fundamentally different. This religious and intellectual divergence was heightened by an elitist limitation of higher education in disciplines such as astronomy or medicine. Medieval Muslim students in what today we call the sciences received instruction almost entirely outside of the regular educational system, usually at an institution sponsored by a princely court, or from individual, often court-supported, scholars.

The pervasive influence of Islam as Faith set the central course of much Muslim learning, as it does today. In education, as in other aspects of life, the strictly traditional view held that the only purpose of thought was to lead one in the right path to the right goal, which was the proper life in service to God's will. In this view, knowledge for the sake of knowledge was not only frivolous but dangerous, possibly heretical. Insofar as innovation departed from traditional thought and application, it invited rejection. Imitation of anything alien was to be scorned: Had not the Prophet warned that anyone who imitates a people becomes one of them? If the people who were imitated happened to be infidel, that is, non-Muslims, so much the worse.

Quite a few Western, non-Muslim historians of science, writing in the earlier decades of the twentieth century, described the eleventh century as witnessing the beginnings of a definite decline in Muslim intellectual life, especially in the sciences. This decline has often been attributed to the increasing rigidity of Muslim religious leaders in interpreting the sacred law and in promoting the various Qurʾanic sciences and curricula over the philosophical ones so fervently attacked by theological and mystical thinkers, such as al-Ghazali and his followers. Within this theocratic-versus-secular framework, Islamic civilization is portrayed as having taken a long and steady downward course into broad, cultural stagnation, just as the world of Western Europe was moving dynamically toward its Renaissance.

In the half-century since the end of the Second World War, thanks in good part to the discovery and examination of manuscripts never before known about or published, the broad picture revealed by historians of science and other scholars involved with Islamic studies has evolved significantly. There is general agreement that decline did occur, but this development is now placed later than the eleventh century, perhaps as late as the fifteenth or sixteenth. It is now established that, between the twelfth and thirteenth centuries alone, important work in astronomy and mathematics was completed by al-Tusi and Ibn al-Shatir, in optics by al-Farisi, in pharmacology by Ibn al-Baytar, and in medicine by Ibn al-Nafis, to cite only some of the more prominent examples. Thus, three centuries after the first great flowering of science in Muslim lands, there occurred another, perhaps in some ways even greater, golden age.

It should be kept in mind that comprehensive and meticulously documented research in the sciences of medieval Islam is a relatively recent endeavor, generated largely in the twentieth century. Hundreds, perhaps thousands, of manuscripts produced between five hundred and one thousand years ago remain in archives around the world, awaiting the resources that will allow scholars to fully study, translate, and publish them, thus helping to clarify and broaden today's knowledge of historic scientific achievement.

The complex, sometimes contradictory character of the changes in Islamic scientific endeavor has come to be perceived as reflecting a process of acquisition, a profound adaptation, and, in a sense, conversion. The middle and late centuries of the medieval period in Muslim lands witnessed a gradual evolution of the sciences that had been imported by farsighted Muslim rulers and first made available in translations largely undertaken by non-Muslims. The Hellenistic character of the texts was long reflected in the character of the teaching that utilized them. In time, however, new generations of scholars in all parts of the Muslim world, shaped by indigenous Islamic traditions, both religious and philosophical, inevitably gave science a predominantly Islamic character. This trend did not spell the total eclipse of the rational sciences. Their value had been fully recognized: Mathematical astronomy had become vital to religious practice; arithmetic and algebra were central to the workings of *fara'id*, the legally determined division of estates; logic became an essential element in the analytical and explanatory religious sciences. Moreover, many *madrasa* became increasingly hospitable to the rational sciences, especially after the advent of a theologically and philosophically inspired dissociation between astronomers and mathematicians on the one hand and astrologers on the other.

The decline in late medieval Islam may be thought to have accompanied the process of Islamicization of the sciences—a long development, in which the Hellenistic approach to knowledge gave way to an Islamic view in which the sciences are assigned to serve as practical instruments in lighting the way to salvation as defined in the Islamic Revelation. This development reached beyond simple conflict between reason and faith. It involved complex and sophisticated intellectual accommodation which was theological, philosophical, and rational.

It remains a major challenge to historians to investigate and illuminate the reasons why the vigor of scientific inquiry apparently atrophied among Islam's scholars at different times and in different areas toward the end of the Middle Ages. By the time Islam's last great dynasty, that of the Ottoman Turks, was spreading into Europe, the Near East, and North Africa, science in Muslim lands appears to have become largely a matter of collecting already collected knowledge and copying it down again and again with minor variation. Original texts seem to have become fewer; *ijtihad*, personal and original intellectual effort, seems to have languished. Some of the giants among late medieval Muslim thinkers were affected by theological controversy. The philosopher Ibn Rushd was the most eminent victim: he, like Socrates long before him and Galileo later on, joined the heroic roster of those ostracized, persecuted, or imprisoned because of their beliefs. Such repression of thought, of course, has occurred in most societies throughout history.

Medieval Islam's isolation from societies outside its borders acted as a parochial brake on its exchange of ideas and techniques, especially after Europe had begun to emerge from the Middle Ages with a new, worldly, inquiring spirit. Muslim trade had dominated the Mediterranean and Near East for centuries, but Islam's traders had never penetrated Europe permanently on any great scale. Muslims remained generally uninterested in Europe almost until the beginning of the modern era: What could the land of the infidel offer beyond profitable commerce? On the other hand, European traders visited Islamic lands in increasing numbers after the tenth century, and Europeans became aware of the things Muslim lands offered: new foods, new materials, new ways of making things and of living, let alone the store of classical learning appropriated, enhanced, and extended by Islamic philosophers and scientists.

By the time Muslims began to investigate the West and then to import some of its ideas, procedures, and practices, the West was immersed in the fundamentally secular and humanistic Renaissance. The rulers and royal deputies of Ottoman Turkey, which at the end of the seventeenth century occupied about half of the regions bordering the northeastern, eastern, and southern shores of the Mediterranean, were forced to take notice of those Western innovations they needed for their own survival. Most of these had to do with the arts of war and with the techniques of administrating an imperial bureaucracy. The military conquests of the empire were largely finished; increasingly caught up in competition and conflict with the growing economic and political power of Europe, the Ottomans were forced to face the huge tasks of modernizing their own civil society.

Only in Southeast Asia did Islam expand, and that primarily in connection with sea-borne trade and commercial exploitation. Islam as Empire was headed for disintegration, much of its territory destined to fall under European colonial domination. There would be further cultural atrophy, revolution, then ultimate evolution into modern, independent states, whose boundaries bore little relationship to anything beyond European geopolitical maneuvering. Through it all Islam as Faith was destined to endure and expand, ultimately embracing one out of every five persons on earth. Today it is clear that Islam as Civilization has endured, as well. And in the last half-century it has manifested a strong new spiritual, political, and cultural vigor.

What has become of Islam's historic scientific achievement? What the

Muslims appropriated and transformed has, of course, become an integral part of an international, world science. Without the Islamic legacy the development of what we know as the modern world would have been significantly different and, at the very least, would have taken much longer, so that we would today be living in many ways in a significantly earlier time. And modern science would be noticeably less than five hundred years old at the end of the twentieth century.

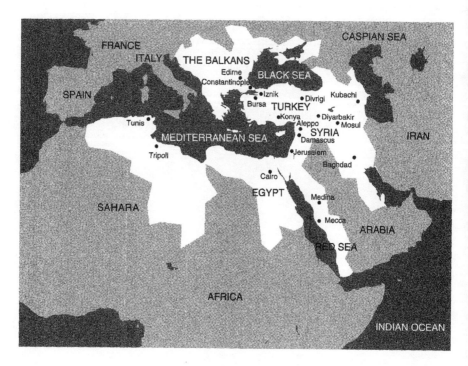

Figure 13.1
Map of Islam in the Late Eighteenth Century

Beginning in the fourteenth century, Islam as Empire suffered military, political, and cultural decline, a trend that was not to be reversed until modern times. Conquest by Seljuk Turks, the Norman invasion of Sicily, Christian crusades in the holy lands, as well as the reconquest of Muslim Spain—all brought upheaval and instability to Islamic domains, with a consequent loss of social vitality and purpose. Among other enterprises, after the fifteenth century science in Muslim lands became eclipsed by the innovative energy and achievement being developed in Western Europe. By the beginning of the eighteenth century the last great Islamic imperial dynasty, that of the Ottoman Turks, occupying about half of the regions bordering the northeastern, eastern, and southern shores of the Mediterranean, had turned in upon itself, occupied increasingly with self-defense against resurgent West European forces that had been revitalized by Renaissance spirit and advance. Now Muslims were obliged as never before to learn about Western innovations they needed for their own survival, such as the arts of war and the techniques of administrating an imperial bureaucracy.

14

Transmission

The widespread and rapid translation of Greek philosophical and scientific works into Arabic following the Muslim conquests of the seventh and eighth centuries had barely passed its peak before the next great wave of transmission began. Around the twelfth century, teams of scholars began the translation into Latin of the considerable and enhanced treasury of Muslim science, that is, science in Arabic. Much of this work was done in Spain, the principal cultural crossroads joining medieval Islam and Europe. This enterprise, for the first time making Greco-Arabic knowledge available to scholars and scientists in Western Europe, accelerated considerably after the Christian reconquest of Spain had embraced Barcelona, Toledo, and Seville.

The translators included Jews and Christians from Spain and other parts of Europe, as well as Muslims. This multiracial, international fellowship remained hard at work until Jews and Muslims were expelled from Spain in the late fifteenth century. *Mozarab* (Christians living under Muslim rule as if they were Arabs), *Mudejar* (Arabs remaining in Spain after the Reconquest), and *Moriscos* (Muslims who, outwardly at least, converted to Christianity) also played a part in the undertaking, which produced what was in effect a twelfth-century pre-Renaissance renaissance that provided the libraries of Europe's monasteries and other intellectual centers with a flood of classical knowledge never before available in the West and now enriched with Muslim contributions.

Among these translators, the most famous by far were an Italian, Gerard of Cremona, a Scot working in Sicily, Michael Scotus, and a Carthaginian, Constantine the African. They, their contemporaries, and their successors provided translations of the major Greek scientists, from Euclid and Hippocrates to Ptolemy, Galen, and others, as revised and annotated by such Muslim thinkers as al-Razi, al-Battani, Ibn Sina, and al-Khwarizmi. Gerard was especially prolific, turning out more than one hundred translated texts. The range embraced by the translators was comprehensive, including medicine, the natural world, meteorology, geology, mathematics, and physics. With this Muslim-enriched legacy, Europe's scholars were inspired to revolutionize their view of the world around them, as well as to redefine current intellectual, technological, and social goals. Medicine and the physical sciences received particular attention, as they had in the first flood of translation, reflecting renewed emphasis on practical needs. Aristotle's scientific texts were fully translated before some of his more abstract philosophical works.

The work of translation on such a scale involved many frustrations, hazards, and deficiencies. The Arabic language presented medieval translators (whose facility with the original Greek was generally minimal) with a truly difficult obstacle course, filled with alien linguistic devices, as well as terms that defied translation and thus had to be simply transliterated. Identification and selection of works to be translated often presented difficulties, since many Arabic manuscripts represented not only the work of different translators but even of different authors not always clearly identified. Sometimes it was hard to distinguish between what was original Greek content and what was subsequent commentary by someone other than the original author. Moreover, since many translators found Arabic easier to handle than classical Greek, a considerable number of original Greek works were given less priority or overlooked entirely.

Despite the difficulties of translation and the sometimes erratic selection of works to be translated, the great majority of important ancient texts were available to European readers of Latin by the fourteenth century. Soon came the means of disseminating such a wealth on an unprecedented scale: In 1445, Johann Gutenberg printed his first book from movable type.

Invigorated with a new, humanistic spirit and armed as never before with a classical heritage, the Western establishment, focusing on concerns closer to home, soon abandoned much of its interest in Islam and forgot the significant Muslim contribution to its culture. At the end of the Middle Ages the modern, and far from trusting, relationship between the West and Islam was beginning to take shape.

The cultural influence of Islamic civilization on the Christian West affected everything from architecture to zoological research. The influence of Muslim sciences was perhaps less apparent than that of Muslim arts. However, few if any of the scientific disciplines that began to be transformed in the West during the late medieval and early Renaissance centuries could have developed as they did without the clarification, renovation, and enhancement that had been achieved by Muslim scientists.

By the seventeenth century, when Ottoman Turkey gave up conquest and began losing control of some of its territory, Europe was on the offensive, militarily, economically, and culturally. New technological instruments and techniques had been developed or imported: papermaking, printing, and gunpowder had begun to change the way whole populations lived, battled, perished, learned, and survived. New social and cultural forces were shaping a civilization that could only become progressively more alien to an Islamic one increasingly less dynamic in character, form, and purpose. The growth of a prosperous and influential merchant class; the Italian Renaissance with its secular, neo-classical, anti-clerical interests; the fierce Reformation that split the Christian Church—most of these developments could only be perceived by Muslims as threatening.

Furthermore, by the nineteenth century, when Islamic lands had begun to fall under colonial control by European powers, devout Muslims could only perceive as a particular menace the Western secularism generated by increasingly humanistic ways of thought focused far more on humans on earth than on God in heaven. The remnants of the Muslim empire, losing unity, strength, and supremacy, settled mostly into a defensive and restrictive political and cultural rigidity. Islam as a culture was closing its doors and its mind to the dangerous world taking shape just beyond its shrinking borders. Many decades would pass before much new and significant knowledge could be imported into Islamic society from outside, let alone generated within it.

The growing dominance of Europe ultimately brought about cracks in the Muslim wall. Ignoring traditional warnings about imitating foreigners, Ottoman leaders adopted some practical Western developments that had clearly promoted the growth of European power, innovations such as new military and naval weapons and strategies and the establishment of a postal system that could help to disseminate necessary knowledge about what was afoot in the world beyond Ottoman borders. The Ottomans also promoted book publication on an unprecedented scale. Moreover, Muslims began going abroad as never before to scout the new world and its ways, and increasing numbers of Western visitors were admitted to Muslim territories. However, the cautionary wisdom that prevailed in Islam for generations urged Muslims not to get too close to Westerners but to learn what they could about them. As the world proceeded into the modern age some understanding developed between Muslims and infidels; this was to increase at times, decrease at others, but the gains to both sides accrued slowly. Meanwhile, the revitalized West took its turn at world leadership and domination.

Figures 14.1a, 14.1b, and 14.1c
Required Reading for Western Europe

The torrent of translation of Greek and other classical knowledge into Arabic that greatly expanded the development of early Islam's cultural environment had barely run its course before it was followed by another outpouring of translation. This time the movement was from Arabic to Latin, which gradually made available to Western Europe an intellectual heritage shaped by both classical knowledge and by the extensions and additions contributed by Muslim philosophers and scientists. Much of the translation was completed by teams of scholars in Muslim Spain, the most open gateway to the European continent. The translators included Muslims, Christians, and Jews from many geographical areas.

As was seen in the earliest Islamic settlements, medical and other scientific texts were in particular demand by Western communities reawakened to the need for social and economic progress. For nearly half a millennium, works such as the three fifteenth- and sixteenth-century examples illustrated in Figures 14.1a, 14.1b, and 14.1c served, for generations of Western students, as essential texts for the specific disciplines to which they relate: Aristotle's philosophical discourse on the soul, *De Anima* (here with comments by Islam's Averroës [Ibn Rushd]); *Al-Qanun* (The Canon of Medicine) by Avicenna (Ibn Sina); and the *Liber ad Almansorem* (Book of Mansur) by the philosopher–physician Rhazes (al-Razi). Renaissance Europe did its scholarly and professional homework with an enormous and encyclopedic Greco-Arabic library, ultimately available in every European language and spread throughout the continent by the printing press.

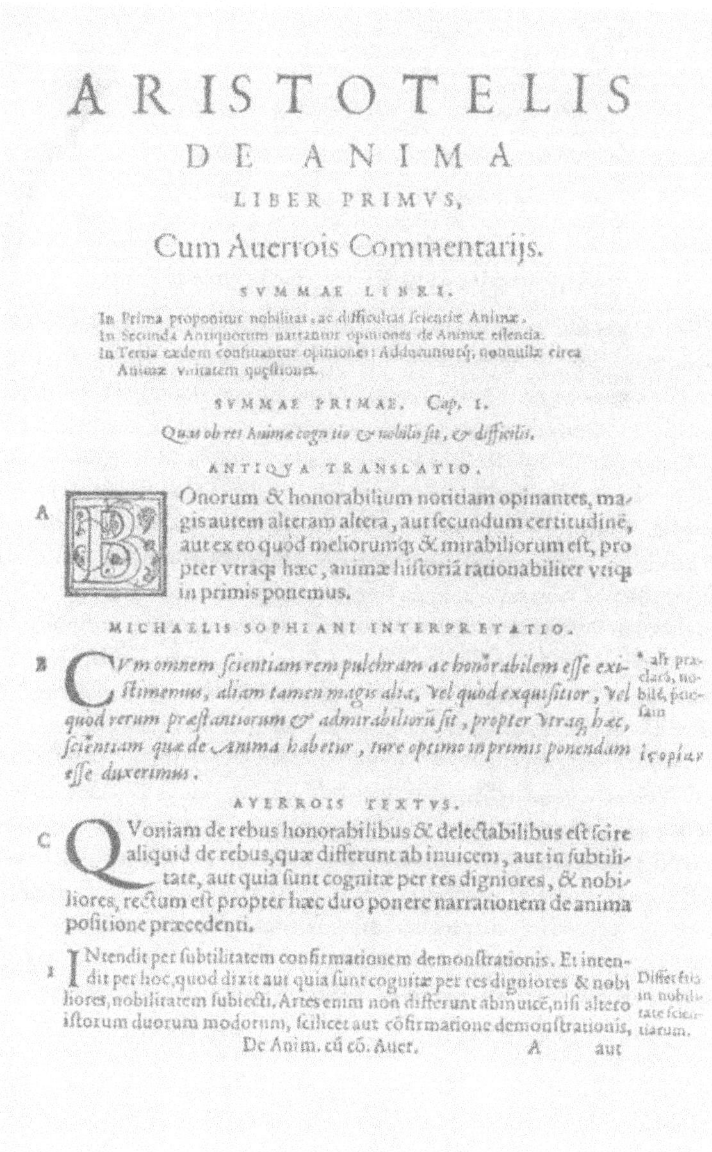

ARISTOTELIS
DE ANIMA
LIBER PRIMVS,

Cum Auerrois Commentarijs.

SVMMAE LIBRI.

In Prima proponitur nobilitas, ac difficultas scientiæ Animæ.
In Secunda Antiquorum narrantur opiniones de Animæ essentia.
In Tertia eædem confutantur opiniones: Adducunturq; nonnullæ circa
Animæ veritatem quæstiones.

SVMMAE PRIMAE. *Cap. I.*

Quas ob res Animæ cognitio & nobilis sit, & difficilis.

ANTIQVA TRANSLATIO.

A **B**Onorum & honorabilium noticiam opinantes, magis autem alteram altera, aut secundum certitudinē, aut ex eo quod meliorumq; & mirabiliorum est, propter vtraq; hæc, animæ historiā rationabiliter vtiq; in primis ponemus.

MICHAELIS SOPHIANI INTERPRETATIO.

B *Cvm omnem scientiam rem pulchram ac honorabilem esse existimemus, aliam tamen magis alia, vel quòd exquisitior, vel quòd rerum præstantiorum & admirabilioris sit, propter vtraq; hæc, scientiam quæ de Anima habetur, iure optimo in primis ponendam esse duxerimus.*

* alt præ-
clarè, no-
bilè, præ-
clarū
ἰςορίαν

AVERROIS TEXTVS.

C **Q**Voniam de rebus honorabilibus & delectabilibus est scire aliquid de rebus, quæ differunt ab inuicem, aut in subtilitate, aut quia sunt cognitæ per res digniores, & nobiliores, rectum est propter hæc duo ponere narrationem de anima positione præcedenti.

1 **I**Ntendit per subtilitatem confirmationem demonstrationis. Et intendit per hoc, quod dixit aut quia sunt cognitæ per res digniores & nobiliores, nobilitatem subiecti. Artes enim non differunt abinuicē, nisi altero istorum duorum modorum, scilicet aut cōfirmatione demonstrationis,
De Anim. cū cō. Auer. A aut

Differētia
in nobili-
tate scien-
tiarum.

Figure 14.1a
Title Page in Latin, from Sixteenth-century Printed Copy
of Aristotle's *De Anima*, with Commentary by Averroës

[214]

TRANSMISSION

Quartus canonis Auicēne cum preclara Gentilis fulginatis expositione.
Thadei itez florentini expositio super secunda Fen eiusdem.
Gentilis florentini iterum super duos primos tracta.quinte Fen.

Quintus etiam can.cum eiusdez Gentilis fulgi. lucidissima expositione.
Canticorum Liber cum cōmento Auer.

Omnia accuratissime reuisa atqz castigata:ac quantum ars anniti potuit fideliter impressa.

Figure 14.1b
Page from Sixteenth-century Latin Translation of Avicenna's
The Canon of Medicine

Figure 14.1c
Page from *Liber ad Almansorem* by al-Razi, Printed in Latin, Fifteenth Century

15

The New West

The full emergence of the experimental method was perhaps the most revolutionary development in the history of science. Its time and place were opportune: the sixteenth century and Western Europe. The static, well-ordered establishment of church and kingdom that had kept all elements of society north of the Pyrenees and west of the Oder in fixed stations and on predetermined courses for long periods in the centuries after the breakup of imperial Rome was now under siege, confronted by the new spirit of a new era—the Renaissance.

One of the elements inspiring this historic rebirth was renewed, humanistic confidence in each person's intrinsic worth and potential—an echo of the central temper of ancient Greece. This was accompanied by a reexamination of the universe in which human beings were finding a more important moral and social position, and by a reappraisal of the role of the Church in human affairs. Scholasticism, the "official" philosophy promoted by Catholic authorities, had attained its zenith around the thirteenth century. But growing conflict between popes and emperors, excesses of the Inquisition, the Church's dread tribunal, and reawakened interest in the philosophy of the ancients, notably the Greeks, all helped to promote schism between proponents of the supremacy of faith and those advocating the sovereignty of reason. Dante Alighieri, the greatest poetic voice of the Middle Ages and last of the great medieval thinkers, had, in his *Divine Comedy*, assigned philosophers such as Aristotle and Averroës to Limbo. The Renaissance rescued most of these figures, setting them up in the kind of lofty, classical pantheon displayed by Raphael in his Vatican fresco, *The School of Athens*.

The most profound shift in human thought attending the birth of the new age involved the revitalized ability of thinking individuals to separate their rational, in-the-street, at-work concerns from their religious devotions. This breach between reason and faith could only force the Church toward generally defensive positions in matters of doctrine, which could not accommodate serious conflict for long. The spreading reliance in the Western world on rational means of intellectual exploration—scientific inquiry—represented a threat and provoked confrontation that has lasted into modern times. If in the past five centuries faith has occasionally claimed its victories, so also has reason, and the struggle seems eternal. Between the sixteenth and twentieth centuries, however, the framework of world science became increasingly secular, with most widely accepted processes of inquiry and confirmation based largely on rational rules.

Not that mysticism, superstition, and magic, let alone religion, disappeared from the scientific community in that period. For example, many of the most eminent scientists of the Renaissance, such as Copernicus, Tycho Brahe, Kepler, Sir Isaac Newton, and René Descartes were involved in astrology, even as their work was undermining its doctrines. The alchemist theories of the Swiss physician Philippus Aureolus Paracelsus aroused widespread interest. The historical record of such activities, together with the more important and sustained influence of religious belief, has tended to challenge the nineteenth-century view of science as a purely positivist endeavor, based exclusively on the empirical analysis and verification of observable phenomena. Today there appears to be significant, perhaps growing, belief in both spiritual force and moral purpose behind scientific endeavor, and this is no longer limited to followers of specific religious dogma.

The experimental method, with its dependence on rational and systematic procedures of inquiry and proof, was not, as we have seen, unknown to medieval scientists. The optical researches of Ibn al-Haytham reflected his notably empirical bent. In the thirteenth century the British philosopher Roger Bacon ventured beyond deductive methods of scholasticism to emphasize the application to scientific inquiry of mathematics and the careful observation of natural phenomena, as well as the validation of results. Three centuries later his fellow-countryman and namesake, Francis Bacon, a philosopher and essayist who was England's lord chancellor under King James I, promoted rational and empirical methods of scientific investigation which he felt could alleviate human misery. It took a few more centuries before Bacon's methods were widely applied. However, they and the experimental philosophy they served were at the heart of the Scientific Revolution that came to dominate international scientific endeavor after the sixteenth century.

Just as the Renaissance built upon and reshaped the medieval civilization nurtured by church and imperial dynasty, so the Scientific Revolution built upon and transformed the Greco-Islamic sciences that had been nourished in medieval Islam and inherited by the Christian West. From the tenth century on, Europe had gradually come to perceive, envy, welcome, and exploit its scientific legacy from Islam, thanks to increasing cultural traffic with Muslim lands via the busy Spanish and Sicilian gateways, the thriving routes of Mediterranean and overland commerce, and the contacts left over from the Crusades. The prestige of Muslim science had reached its highest level in Europe around the twelfth century, just before Muslim scientific vigor began to fade in some of its places of origin. But Europe's admiration for Islam's scientific gift became overshadowed by European envy and fear of what was long perceived as Muslim military and political supremacy, and by Christian hostility toward a faith that appeared false, violently aggressive, and self-indulgent. Intellectual contact between the two civilizations had reached new heights; now political and religious enmity between them was intensifying. In addition, in the West the breach between religion and science was widening. These parallel developments were to affect profoundly the very different ways in which science was to develop from that point in both the Western and Muslim lands.

The revival of interest in human beings' potential and purpose that marked all of the great movements, trends, styles, and diversions of the Renaissance and post-Renaissance centuries, and that have shaped much of our own cultural foundation in the Western world, did not signify a total revolution in philosophy. The Middle Ages in Europe had long been dominated by an unending conflict between Church dogma and a kind of humanistic and individual quest for intellectual liberation. However, a crucial condition shaped post-medieval civilization in the West and distinguished it from the civilization of Islam: in the West, the Church was losing the battle, while in the Islamic lands orthodox religious authority was further consolidating its hold. The great schism in Europe that produced the Protestant sects did not long arrest the growing separation of church and state nor seriously delay the radical and often anticlerical transformation of virtually all societies in the West.

Thinkers and scholars of late medieval Europe became greatly influenced by the spread of scientific ideas imported from Islam, including the concepts of natural philosophy that were embedded in the enhanced Greco-Islamic intellectual legacy. Elements of Aristotelian and Platonic thought, often filtered through Muslim interpretations and emendations, combined with Christian concepts to provide powerful inspiration for a broad range of philosophical and scientific investigation involving a variety of disciplines, theological, metaphysical, mathematical, and medical. This vigorous intellectual activity was sponsored in good part by the great universities and colleges established in Europe by the end of the Middle Ages. Here was a growing and changing institutional resource on a scale unknown anywhere before.

After the seventeenth century, one intellectual development after another tended to restrict the domain of religious rule and expand the reach of civil authority in the West. The intellectual freedom of the Enlightenment and the philosophical individualism embodied in Romanticism promoted radical beliefs in individual liberty, in equality, and in what came to be accepted as the democratic political rights of each human being, inspiring intellectuals and political activists throughout the West to focus on revolution as essential to political advance and social progress. Individual ways of thought and action were inevitably altered by the revolutionary currents sweeping across the European continent. Increasing reliance was placed on individual viewpoint, individual judgment, and on a most liberal and radical notion: one must find one's own ways of dealing with God.

Through the four centuries that followed the emergence of the Renaissance—through the Reformation and the Counter-Reformation, revolution and counter-revolution, through decades of war and peace, destruction and progress, the civilization of the West took modern shape. And if the last half-millennium has come to be noted for Western achievement in the arts, it has come to be even more identified with Western science, deeply embedded in which is the Greco-Islamic legacy inherited by the West from the civilization of Islam. It is ironic that this Muslim gift played a part, from the Renaissance on, in promoting a humanistic scientific spirit that would increasingly bypass or ignore religion. At the same time the Islamic world, whose greatest medieval scientists had notably resisted the rigid proscriptions and narrow strictures of the *ulema*, was moving toward an increasingly theologically enclosed culture that, given its geopolitical setting, no longer could promote vigorous or original scientific effort. The cultural dialogue between East and West, beginning to expand at the start of the modern era five hundred years ago, became severely diminished, with consequences that are still shaping the world today.

16

Epilogue

The Safavid and Qajar empires in Persia, and the Ottoman empire in Turkey, the Fertile Crescent, and Egypt, were the last of the great Muslim dynasties to rule most of the lands conquered by Islam in its first two centuries. The greatest of these powers were the Ottomans, who were first in the post-Renaissance era to impose a kind of modern colonialism on the various Muslim peoples living around the Mediterranean basin and in the Near East. Ottoman rule, established in the fourteenth century, remained unvanquished if not unchallenged for more than five hundred years. In the sixteenth century the greatest Ottoman ruler, Sulayman the Magnificent, established an essentially Turkish central government, which achieved a well-organized, sophisticated administration that left permanent marks on every region under its control.

By the nineteenth century, political and economic reforms, many adopted through increasing study of European communities, were generating an increasingly modern society. Now at last, Muslims, long turned inward and away from "contamination" by the West, began to observe many of Europe's governmental and educational procedures. Travel and communication, long predominantly West-to-East, became more reciprocal. The imbalance established so long before never shifted significantly, however: Westerners remained far more interested in Islam's civilization than Muslims were in the world beyond Islam. Finally acknowledging the West's military, spiritual, and political threat, Muslims felt obliged to study Western institutions and ways. However, their investigative efforts came too late to enable them to begin competing fully in economic or military terms until after the discovery of Middle East oil in the early twentieth century.

Between the mid-eighteenth and the twentieth centuries, Ottoman domination of Muslim territories was replaced by British, French, Russian, and Italian rule over most of the same lands, save Turkey itself. No Muslim Arab land gained full independence until after the First World War. For more than a century Muslims watched as their leaders imported whichever Western technologies, educational procedures, and political strategies appeared to offer acceptable ways of modernization and effective commercial competition with the world beyond Islam. The results of such large-scale adoption of often alien concepts and methods ranged from profound to superficial, from beneficial to brutal. Probably the most radical changes were achieved in Turkey in the 1920s and 1930s under the rule of the secular revolutionary, Kemal Atatürk.

Since the beginning of the Second World War, traumatic changes have continued to shape virtually all Muslim countries. Yet prewar colonial geopolitics have left their mark, most apparent in the national boundaries drawn up as a form of diplomatic insurance by departing colonial rulers. The colonial legacy has had additional, deeper effects: most influential has been the shunting aside of traditional Muslim legal systems, standards of ethics, and social conventions in order to accommodate Western ways, many of which have been felt by many Muslims to be alien to their culture. It is not surprising that the last half century has witnessed the emergence of a desire to return to fundamental Islamic traditions. The spiritual, ethical, and social ways of the past have come to be regarded with increasing respect by more and more Muslims who, in a world that they have perceived as largely hostile or indifferent, have come to feel the need for a more assertive Islamic identity and the feelings of security that it may promise.

New observations of star positions made in the fifteenth century at the Samarkand observatory established by the Mongol ruler Ulugh Beg were among the last and most important achievements of medieval and post-medieval Muslim scientists. Much subsequent science in Islamic lands appears eclipsed by Western achievement, on which the Muslim world long depended for much of the practical technology needed for survival in the modern world. Not until the end of the Second World War did Islamic countries stir themselves out of the cultural miasma of nearly four centuries, emerging into a world which they have generally accommodated with as much difficulty as the Western world has encountered in accommodating them.

Vanished today is Islam as Empire; still extant, if sometimes obscured, is the enduring Islamic blend of unity of belief and diversity of culture. The Islamic faith not only remains fundamentally strong, it is gaining adherents. It remains the firm core of Islamic unity, except for the intermittent resurgence, here and there, of dogmatic, fundamentalist extremism. As for Islamic culture, its vitality endures today with a broader than ever geographic and ethnic diversity that reflects the varied lives and indigenous customs of hundreds of millions of Muslims spread not only between Morocco and Indonesia but also between the Atlantic and Pacific.

Islam as Community? The discovery and exploitation of the vast oil reserves of the Middle East have propelled Muslim nations—especially the Gulf states—into great (and uneven) wealth, worldwide commercial influence and rivalry, as well as intermittent political and social discord. In addition, the birth of the independent state of Israel amid territory held by most Muslims to be traditional homeland; the wars resulting therefrom; the problem of the Palestinian people which each Arab and non-Arab nation concerned has tended to weigh in terms of its own needs and strategies—all these factors have congested the historical landscape, obscured the horizon, and delayed fatefully the development of practical and just resolutions to major problems afflicting Muslims and non-Muslims alike.

Confounding the Muslim situation and profoundly affecting Muslim attitudes is the fact that so many of the most obvious influences on Islamic life in the past century have come from outside—from communities of Westerners, the infidels. It has taken only a very few decades for Islam to turn the most important of these influences into large-scale benefit. The spectacular technological and industrial infrastructure created in petroleum-rich Saudi Arabia and its neighboring states testifies to the Muslim ability for ultimately taking from outside whatever technology can usefully be adapted to Muslim purpose. In this way the science of the modern world, including its Greco-Arabic heritage, has indeed returned to the region where much of it was formed. However, nowhere else could it be more apparent that massive and sophisticated urban engineering complexes, intricate chemical processes, and electronic networks that cover the whole region and connect it with other areas of high technology around the world all represent a world science that can never be categorized in terms of nation, ethnicity, or religion.

The earliest examples of late-twentieth-century architecture to arise in Saudi Arabia—university, medical, and governmental structures—clearly reflect the dramatic unadorned styles made popular by Western architects. In recent years, however, a conscious effort has been made, most notably by the Aga Khan Award for Architecture, to devise a public and private architecture that, while fully meeting the practical and technological needs of today, also reflects traditional Islamic life, with its focus on privacy, family, and communal consensus. The most distinguished of these structures demonstrate an Islamic spirit that goes deeper than surface decoration. This "return" to something akin to traditional qualities has been proceeding successfully. Architecture and perhaps others of the arts can thus be Islamicized to some extent. But in other areas of life the attempt to return to fundamentals, increasingly evident in many Islamic lands today, is bringing about both controversy and resistance.

Much of the opposition arises from the conflict between Muslim appreciation of the efficiency and usefulness of Western methods and products and Muslim perception of the threats Western ideas and moderate or "liberal" Muslim attitudes pose to traditional and orthodox Islamic values. Often this boils down to a conflict between the Muslim faith and secular concepts of Western philosophy and ways of living, and it extends beyond social customs such as the sequestration and veiling of women. The traditional Islamic fundamentalist must still shun the idea of a separation between faith and reason, a distinction long generally tolerated or accepted in most Western societies. In any community that has maintained for generations a static, inward-looking, and exclusive social milieu largely controlled by rigid, pro-forma theology, the best of what the outside world offers cannot be welcomed easily if at all unless it can be utilized without significant deviation from orthodox religious belief. In an Islamic community marked by resurgent fundamentalism, such an attitude threatens once again to become an important and encompassing factor in daily life.

Modern Muslim attitudes toward the character and purpose of science include a broad range of opposition, with an equivalent range of intensity: some would restore what they consider a strictly pure Islamic science—science pursued only as it serves God's purpose as long ago defined by the medieval Islam's *ulema* and their fellow orthodox thinkers. Such aims, long obscured in the Muslims' process of Westernization, are being promoted again, on a scale difficult at present to assess. In one esoteric view that seems either unrealistic or irrelevant (even, to some, ridiculous) to most scientifically oriented Western minds, the finite, geocentric Ptolemaic universe is seen as better serving God's—and thus Islam's—purpose than does the Copernican model, which replaced it more than half a millennium ago.

Thus a metaphysical and theological purpose dominates the discussion of physical and rationally descriptive concepts, which are denigrated in the process. Is this sort of "return to fundamentals" totally unfamiliar? Think closer to home and our own times: recall the cosmology of our Creationists, with their brief timetable as to universal beginnings, and remember also the anti-evolutionist furor that began a hundred and fifty years ago.

There are examples in every culture of non-scientific opposition to scientifically accepted findings. Less doctrinaire Muslim opposition to modern science includes those who would redefine tenets of the faith in order to accommodate the intellectual and technical requirements of present-day society, this reinterpretation being derived from what are seen as the most revolutionary and most liberal precepts included in the revelation received by Muhammad. However, it seems likely that those who strongly oppose the rational principles and purposes of modern science are outnumbered today by Muslims who, to some degree, regard the requirements of their faith and those of daily political and economic life as practically separate, if not in significant conflict. Rendering unto God and also unto Caesar has parallels far from Rome.

Of course, the confrontation between faith and reason does not end. Throughout history faith has overcome reason, faith has spurred reason, reason has denied faith, reason has made place for faith. In that sense, the history of Islamic science is not radically different from the history of science beyond Islam. The list of embattled scientific genius is a world list, not a parochial or regional one: Ibn Sina, Galileo, Charles Darwin . . . . Legions of scientists have managed to resist, not the demands of faith, but the rigid servants, or henchmen, of faith who interpret and apply it in oppressive ways. This is a struggle between human beings and it is never finally to be won nor finally lost.

Some scientists, as well as historians of science, recognizing the historic Islamic contribution to the development of world science, and aware, too, of the historical eclipse in Muslim scientific pre-eminence by Western advances, express the hope that science in Islam will experience its own renaissance, especially now that Islam is again asserting its importance as a major spiritual and political force in the world community. There is wide recognition of the need in many Muslim countries for expanded scientific training, research, publication, of exchange of findings with non-Muslim scientific establishments. Such a renaissance may well develop within a framework of Muslim religious belief, adapting for its own purposes the kind of experimental philosophy that still dominates science in Western and other non-Islamic societies. In any event, in the West the rationalism that has long governed scientific thinking is already being increasingly informed by moral or ethical questions, especially in connection with the ecological and social effects of today's vast scientific enterprise. What price our modern sciences? What are their aims? What and whose purposes do they ultimately serve?

Moreover, is our search for knowledge about the universe we live in adequately served by what we have long considered the scientific method of inquiry? Can we in the West be helped by the current trends toward reconnecting pure science to ethics or religious beliefs? How certain are we that our systems of science can adequately explain the universe?

As the twentieth century draws to a close, it appears that fundamental questions concerning the relationship between science and ethics are being raised with an accelerating urgency, in connection with the proliferation of nuclear weapons as well as the advances in genetic engineering, to name only two matters currently arousing international argument. Underlying the thinking applied to dealing with such major issues is the need to work toward truth: material truth together with moral truth. This inevitably involves the continuing encounter between those who consider the intellect as the source of all truth and those who can accept only religious or spiritual revelation as that source. This issue has occupied both Eastern and Western civilizations for centuries. In many Western minds, it has been largely won, in modern times, by warriors armed with intellect. The contemporary Islamic world reflects a different condition: secularism has influenced many of the outward aspects of daily life, as well as considerable areas of thought and belief. At the same time Muslim life continues to be profoundly nourished and guided by the demanding message of the Islamic Revelation. Islam may have been extensively secularized in our century, but it has not become a secular civilization. The confrontation between reasoning and revelation as sources of ultimate truth remains active throughout the worldwide Islamic community, with an intensity generally unmatched in the West.

More than any other of the world's best-known sacred texts, the Qur'an includes references to human responsibility for understanding and preserving the earth and all its life. These are concerns shared by most of the world's people regardless of creed. Even those who are convinced that the world's environmental and social problems can be fully dealt with through logic and rational experiment alone will not be the losers by opening their minds to other, differently oriented, approaches. Whatever the road chosen, the destination will be shared by all. In that sense, as concerns the progress of science, let alone that of all other human enterprise, the Islamic world and the non-Islamic world must understand more fully the roads each has traveled thus far, or they will get lost on the road ahead.

Figure 16.1
(on facing page)
Map Showing the Peoples of Islam Today

The Southeast Asian nations of Bangladesh, Indonesia, and Malaysia today contain almost as many Muslims (more than 200 million) as do the historic Islamic regions of the Near East and North Africa. The balance of the world's present Muslim population is spread throughout all the continents. All in all there are more than 935 million Muslims, nearly one-fifth of all earth's people.

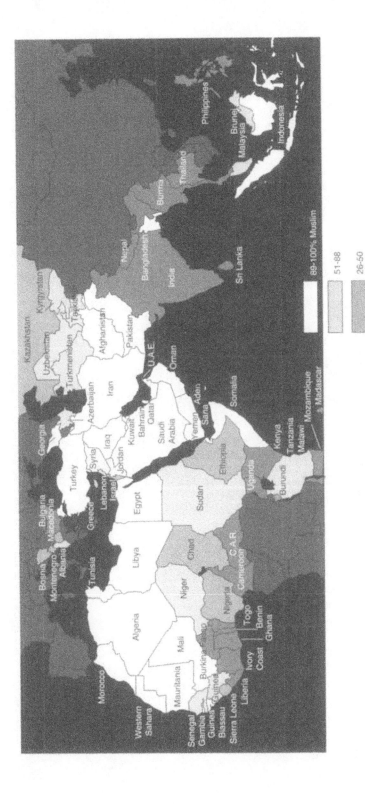

Islam and the World
A SUMMARY TIMELINE

Muslim World	Non-Muslim World
1ST CENTURY A.H.	**7TH CENTURY A.D.**
622 Hegira of Muhammad and followers to Mecca; start of Muslim calendar.	600 Rise of the Mayan Empire.
	628 Byzantine victory over Sasanians.
650 Qur'an put in written form.	645 Buddhism reaches Tibet.
659 Muslims split into Shii (Shiite) and Sunni sects.	664 England accepts Roman church.
692 Dome of the Rock completed in Jerusalem.	
698 Arabs conquer North Africa.	
2ND CENTURY A.H.	**8TH CENTURY A.D.**
750 Abbasids destroy Umayyad caliphate and move capital to Baghdad.	700 Decline of Mayan Empire.
	730 Printing invented in China.
760 Arab scholars adopt Indian numerals.	732 At the Battle of Tours, Charles Martel, ruler of the Franks, halts Muslim expansion beyond Pyrenees.

Note: The Timeline on this and the following pages was prepared in 1981 by Ray Graham & Associates for use in The Heritage of Islam Exhibition and has been adapted for this book by the author. Specific dates are given according to the Gregorian calendar.

Muslim World

785 Construction of the Great Mosque of Córdoba begins.

786 Caliphate of Harun al-Rashid begins at Baghdad.

3RD CENTURY A.H.

825 Muslims begin conquest of Sicily.

832 House of Knowledge founded in Baghdad; translation of Greek works into Arabic.

847 Construction of the Great Mosque of Samarra.

851 Earliest Arabic description of China and Indian coast.

4TH CENTURY A.H.

925 Death of Arab physician and philosopher al-Razi (Rhazes).

934 Birth of Persian poet Firdawsi.

ca.935 Qur'an in final written form.

969 Fatimids build Cairo as new capital.

976 Al-Azhar University founded at Cairo.

Non-Muslim World

778 Failure of Roland's expedition against Muslim Spain.

793 Viking-Norman raids begin in Europe.

9TH CENTURY A.D.

800 Charlemagne proclaimed emperor of the Holy Roman Empire.

802 Founding of the Angkor Empire (Cambodia).

820 Normans raid Gaul (France).

853 China produces the first printed book.

878 Alfred the Great defeats the Danes.

10TH CENTURY A.D.

918 Koriyo dynasty unifies Korea.

942 Conversion of Hungary to Christianity begins.

961 Otto the Great becomes Holy Roman emperor.

990 Expansion of the Inca Empire (Peru).

TIMELINE

Muslim World	Non-Muslim World
5TH CENTURY A.H.	**11TH CENTURY A.D.**
1017 Druse sect appears.	1000 Vikings establish colonies in the New World.
1037 Death of philosopher and physician Ibn Sina (Avicenna).	1045 Movable type invented in China.
1050 Kings of Mali convert to Islam.	1054 Breach between Eastern and Western churches.
1055 Seljuk Turks take Baghdad.	
1076 Ghana conquered by Morocco.	1066 Norman conquest of England.
	1085 Christians recapture Muslim Toledo.
	1091 Normans conquer Muslim Sicily.
	1095 First Christian Crusade proclaimed.
6TH CENTURY A.H.	**12TH CENTURY A.D.**
1100 Mathematician-poet Omar Khayyam composes the *Rubayyat*.	1122 Concordat of Worms ends investiture controversy between Papacy and Germany.
1111 Death of Arab theologian al-Ghazali.	1147 Second Crusade begins.
1171 Saladin overthrows Fatimids and founds Ayyubid dynasty.	1190 Third Crusade; Richard the Lion Hearted.
1175 First Muslim Indian empire founded.	1193 Zen Buddhism founded (Japan).
1198 Death of Arab philosopher Ibn Rushd (Averroës).	

Muslim World

7TH CENTURY A.H.

1219 Genghis Khan's first attack on Muslim territories.

1231 Mongol armies attack Persia.

1258 Mongols capture Baghdad.

1260 Mamluks take all of Egypt and Syria.

1280 First Ottoman principality.

1295 Conversion of Persian Mongol ruler to Islam.

8TH CENTURY A.H.

1316 Muslim king in Nubia.

1325 Muslim kingdom in Mali at its height.

1349 Muslim missionaries reach Nigeria.

1377 Death of Arab geographer Ibn Battuta.

1379 Tamerlane (Timur) invades Persia.

1398 Tamerlane invades India.

9TH CENTURY A.H.

1400 Islam reaches Java.

1406 Death of Arab historian and sociologist Ibn Khaldun.

Non-Muslim World

13TH CENTURY A.D.

1202 Mongols conquer Asia.

1215 Signing of Magna Carta.

1226 Death of St. Francis of Assisi.

1271 Marco Polo begins travels.

1291 Eighth and last Crusade ends with defeat of Christians at Acre.

14TH CENTURY A.D.

1309 Papal "captivity" at Avignon begins.

1321 Dante completes *Divine Comedy*.

1337 Beginning of Hundred Years' War between France and England.

1348 Plague ravages Europe.

1370 Aztecs found Tenochtitlan (Mexico City).

1380 Muscovites defeat Mongols.

15TH CENTURY A.D.

1415 English under Henry V defeat French at Agincourt.

1431 Joan of Arc burnt at stake.

TIMELINE

Muslim World   *Non-Muslim World*

1453 Ottomans take Constantinople; end of Byzantine Roman Empire.

1492 Muslim holdings in Spain lost to Christian armies of Ferdinand and Isabella.

1493 Expulsion of Arabs and Jews from Spain.

1445 Gutenberg prints first book from movable type.

1475 Birth of Martin Luther.

1486 Latin translation of al-Razi's *Medical Encyclopedia*.

1492 Columbus discovers New World.

10TH CENTURY A.H.

1502 Shii Islam established as state religion of Safavid Persia.

1520 Reign of Ottoman Sultan Sulayman I (The Magnificent) begins.

1526 Mughal Empire founded (India).

1529 Ottomans repulsed at Vienna.

1529 Islam spreads into Ethiopia.

1535 Ottomans reach Baghdad, Tripolitania, Tunisia, and Algiers.

16TH CENTURY A.D.

1510 First African slaves arrive in the Americas.

1519 Cortèz reaches Mexico and begins conquest of Aztecs.

1519 Magellan starts around globe.

1521 Martin Luther excommunicated; Protestant Reformation begins.

1527 Pizarro begins conquest of Peru.

1588 English defeat Spanish Armada.

11TH CENTURY A.H.

1600 Islam expands in Celebes and Borneo; expansion in Philippines limited by Spain.

1611 Shah Abbas builds new capital at Isfahan, Persia.

17TH CENTURY A.D.

1600 Shakespeare's *Hamlet* written.

1607 First permanent English settlement in America, at Jamestown.

1618 Thirty Years' War begins.

SCIENCE IN MEDIEVAL ISLAM

Muslim World   *Non-Muslim World*

1645 Beginning of Ottoman war with Venice.

1661 Intermittent war between Holy Roman Empire and Ottomans begins.

1683 Ottoman siege of Vienna fails.

1620 Pilgrims reach New England.

1642 Civil war begins in England.

1661 Absolute rule of Louis XIV (the Sun King) begins.

12TH CENTURY A.H.

1707 Mughal Empire begins rapid decline.

1730 Fall of Safavid dynasty.

1735 Wahhabi fundamentalist movement in Arabia sets out to purify Islam.

1770 Russians defeat Ottoman fleet.

1773 Saudi dynasty takes Riyadh; reform movement spreads.

1774 Treaty of Kuchuk Kainarji reduces Ottoman control of the Black Sea.

18TH CENTURY A.D.

1704 *Tales of the Arabian Nights* published in Europe.

1735 Abraham Darby produces iron from coke-fired blast furnace.

1751 French *Encyclopedia* appears.

1767 Watt invents steam engine; Industrial Revolution under way.

1776 American Revolution.

1789 French Revolution.

13TH CENTURY A.H.

1805 Muhammad Ali, Ottoman army officer, establishes rule in Egypt.

1806 Wahhabi reformers take Mecca.

1830 French take over Algiers.

19TH CENTURY A.D.

1804 Napoleon crowned emperor.

1859 First oil well drilled, in the United States.

1859 Darwin's *Origin of Species* published.

1861 American Civil War begins.

TIMELINE

Muslim World

1860 French take over West African Muslim territories.

1869 Inauguration of Suez Canal.

1882 British occupy Egypt.

14TH CENTURY A.H.

1916 Revolt of non-Turkish Muslims against Ottoman Empire.

1918 First World War ends; European powers frustrate Arab independence.

1922 Egypt gains independence.

1923 Mustapha Kemal becomes president of new Republic of Turkey.

1945 Post-war era begins; Arab-Israeli wars; spread of independence to India and to all Muslim states.

1970s Egyptian-Israeli peace treaty. Establishment of Islamic Republic of Iran. Palestine-Israel agreements begin.

Non-Muslim World

1867 Marx' *Das Kapital* published.

1871 The unification of Germany.

1876 Bell invents the telephone.

20TH CENTURY A.D.

1912 Sun Yat-sen proclaims Republic of China.

1914 First World War begins.

1916 Einstein's *General Theory of Relativity* produced.

1917 October Revolution in Russia.

1939 Second World War begins.

1945 United Nations founded; first use of the atom bomb; Second World War ends.

1949 Mao Zedong proclaims the People's Republic of China.

1975 Vatican sets up commission for relations with Islam.

1985 Beginning of disintegration of Soviet Union.

Glossary

al-jabr: The origin of the term "algebra," the Arabic literally denotes a transposition—a restoration of balance or equilibrium—through adding or subtracting the same quantity on both sides of an equation; associated with this activity is reducing or simplifying and combining equivalent terms.

al-Qur'an: Its title literally translated as "recitation" or "reading," this is Islam's holy book.

Almagest (*Al-Majisti*): The astronomical text by the second-century Hellenistic Egyptian Ptolemy, which guided virtually all early Muslim astronomy.

Buraq: The winged steed on which the Prophet Muhammad ascended to the heavens during his Night Journey.

De Materia Medica: A group of medical writings generally attributed to the ancient Greek physician Dioscorides and widely studied throughout medieval Islam and the West.

falasifa: Philosophers who were loyal, in particular, to the principles contained in a Hellenistic version of Aristotelian logic and reason.

fara'id: The legally determined division of estates in countries ruled by Islamic law.

hadith: The written record of actions or sayings attributed to Muhammad.

Hegira: The emigration, or *hijra*, of the Prophet Muhammad from Mecca to Medina.

ijtihad: Personal and original intellectual effort.

Ikhwan-al-Safa': The Brethren of Sincerity, a tenth-century mystic sect that was composed of philosophers who opposed the "orthodox" Aristotelian tenets concerning the nature of matter.

jihad: Although often mistranslated as "holy war," this term is better understood as meaning "effort" or "struggle" (as for the Faith).

kalam: A theology involving dialectical examination unlike the broad kind of secular argument that has been associated with Western theological and philosophical debate since the beginning of intellectual speculation in classical antiquity.

kitab: Book.

madrasa: A mosque school.

mater: A flanged compartment on an astrolabe that holds the *rete*.

mihrab: The prayer niche in a mosque or other Islamic shrine.

miqat: Timekeeping.

muhtasib: Authorities with the rank of judge or legal scholar.

muwaqqit: The official timekeepers employed by mosques.

qanat: A type of underground conduit that links together a series of wells and is capable of tapping distant sources of ground water.

qibla: The direction of the Ka'ba, the central place of devotion at Mecca.

rete: A rotatable grid on an astrolabe, bearing pointers representing prominent stars and a circle representing the ecliptic, or sun's apparent path against the stars.

ruznama: Almanac.

Shariʿa: God's law.

souk: Market.

ulema: A group or community of orthodox Muslim religious leaders.

ummah: The Muslim religious and political community.

zij: Astronomical handbooks with tables providing records of planet and star positions as well as eclipses.

Works Consulted

* Denotes a comprehensive illustrated collection of essays by distinguished scholars surveying, often in considerable detail, the broad range of historic Islamic or Arab achievement in the arts as well as the sciences; also covered are significant developments in religious, political, and social life since the beginning of Islam more than fourteen hundred years ago.

\# Denotes a work which reflects or describes a distinctly Muslim point of view, thus offering the Western reader an opportunity to become acquainted with a very different spiritual, metaphysical, or philosophical, and sometimes esoteric context in which much Muslim life and thought are conducted and expressed.

History of Science

Goldstein, Thomas. *Dawn of Modern Science: From the Arabs to Leonardo da Vinci*. Boston: Houghton Mifflin, 1980.

Lindberg, David C. *The Beginnings of Western Science: The European Scientific Tradition in Philosophical, Religious, and Institutional Context, 600 BC to AD 1450*. Chicago and London: University of Chicago Press, 1992.

Mason, Stephen F. *A History of the Sciences* (new revised edition). New York: Collier Books–Macmillan, 1962. Originally published as *Main Currents of Scientific Thought*. London: Routledge & Kegan Paul, 1953.

Murdoch, John E. *Album of Science*. Vol. 1, *Antiquity and the Middle Ages*. New York: Scribner, 1984.

O'Leary, De Lacy Evans. *How Greek Science Passed to the Arabs*. Chicago: Ares Publishers, 1979.

Price, Derek D. *Science since Babylon*. Enlarged edition. New Haven: Yale University Press, 1975.

Sarton, George. *Introduction to the History of Science* (3 vols. in 5). Melbourne, FL: Krieger Publishing Co., 1927–1948 (reprints). 3 vol ed: Baltimore, Williams and Wilkins, 1927–48.

Sarton, George. *A Guide to the History of Science*. New York: Ronald Press, 1952.

Sarton, George. *The History of Science and the New Humanism*. Midland: Indiana University Press, 1931.
Singer, Charles Joseph. *A Short History of Scientific Ideas to 1900*. New York: Oxford University Press, 1959.
Thorndike, Lynn. *A History of Magic and Experimental Science*. New York: Columbia University Press, 1923–58.

Islam—History and Civilization

Glassé, Cyril. *Concise Encyclopedia of Islam*. With an Introduction by Huston Smith. San Francisco: Harper and Row, 1989.
Lewis, Bernard. *The Middle East: A Brief History of the Last 2,000 Years*. New York: Scribner, 1996.
*Lewis, Bernard, ed., with various scholars. *Islam and the Arab World Faith, People, Culture*. New York, Alfred A. Knopf and American Heritage Publishing Co., 1976.
#Rahman, Fazlur. *Islam*. 2d ed. Chicago: University of Chicago Press, 1979.
Rosenthal, Franz, . *The Classical Heritage in Islam*. London: Routledge & Kegan Paul, 1975.
*Schacht, J. and C. H. Bosworth, eds. *The Legacy of Islam*. 2nd ed. Oxford: Clarendon Press, 1974. Note: the 1st edition (Oxford, 1931) of this work, edited by Sir Thomas Arnold and Alfred Guillaume and authored by an earlier group of scholars contains the same broad range of cultural subjects as does the 2nd edition. Of particular interest is the 1st edition's comprehensive account of Muslim geography and commerce.
Stewart, Desmond, and the editors of Time-Life Books. *Early Islam*. New York: Time Inc, 1967.
Various, *Encyclopedia of Islam*. 2nd (New) Ed, through Vol VI), 1960–; Supplement (Vol 1) New ed., 1982–; 1st ed. (9 vols), 1913–1936; Leiden, E.J. Brill. 2nd ed (1960).

The Arabs—History and Civilization

*Hayes, John R., ed. *The Genius of Arab Civilization*. 2nd ed., rev. Cambridge: MIT Press, 1983.
Hitti, Philip K. *History of the Arabs from the Earliest Times to the Present*. London: Macmillan, 1968.
Hourani, Albert. *A History of the Arab Peoples*. Cambridge, MA: Belknap Press/ Harvard University Press, 1992.
Landau, Rom. *Arab Contributions to Civilization*. San Francisco: American Academy of Asian Studies, 1958.
Nawwab, Ismail I., Peter C. Speers, and Paul F. Hoye, eds. *Aramco and Its World: Arabia and the Middle East*. Dhahran: Aramco, 1980.

Religion

*Ali, Ahmed. *Al-Qur'an: A Contemporary Translation*. Revised definitive edition, third printing, with corrections. Princeton: Princeton University Press, 1990. Note: Muslims consider that only Arabic, the language of the Revelation, can properly transmit the spirit and meaning of their holy book. Other languages can only provide an acceptable paraphrase. The Ali translation is a recent rendering with a modern tone. Earlier English translations that have also found favor among non-Muslim readers include, among others, A. J. Arberry's *The Koran Interpreted* (New York: Macmillan, 1964), and M. M. Pickthall's *The Meaning of the Glorious Koran* (New York: Everyman's Library, Alfred A. Knopf, 1992).

Bucaille, Maurice. *The Bible, the Qur'an, and Science*. Indianapolis: American Trust Publications, 1978.

Geertz, Clifford. *Islam Observed: Religious Development in Morocco and Indonesia*. Chicago: University of Chicago Press, 1971.

Peters, F. E. *Children of Abraham: Judaism, Christianity, Islam*. Princeton: Princeton University Press, 1982.

Islamic Cosmology

#Nasr, Seyyed Hossein. *An Introduction to Islamic Cosmological Doctrines*. Boulder: Shambhala, 1978.

Science in Islam

Anawati, G. "Science." In *Cambridge History of Islam*, edited by P. M. Holt, A. K. Lambton, and B. Lewis, vol. II, pp. 741–79. Cambridge: Cambridge University Press, 1970.

#al-Andalusi, Sa'id. *Science in the Medieval World: "Book of the Categories of Nations."* Translated and edited by Sema'an I. Salem and Alok Kumar. History of Science Series No. 5. Austin: University of Texas Press, 1991.

Gillispie, C. C., ed. *Dictionary of Scientific Biography*. New York: Scribner, 1970–80.

Grunebaum, Gustav E. von, ed. "Muslim World View and Muslim Science." In *Islam, Essays in the Nature and Growth of a Cultural Tradition*. Memoir No. 76. Menasha, WI: American Anthropological Association, 1954.

Hamarneh, Sami K. "The Life Sciences." In *The Genius of Arab Civilization*, 2d edn., edited by John R. Hayes, pp. 173–200. Cambridge: MIT Press, 1983.

#Hoodbhoy, Pervez. *Islam and Science: Religious Orthodoxy and the Battle for Rationality*. London: Zed Books, 1991.

Journal of the History of Arabic Science. Published by the Institute for the History of Arabic Science, University of Aleppo, Syria.

King, David A. "The Islamic Aspects of Islamic Science." Prepared for catalog (unpublished), "Heritage of Islam" Exhibition, 1983.

King, David A. *Catalogue of Scientific Manuscripts in the Egyptian National Library*. Cairo: General Egyptian Book Organization, in collaboration with the American Research Center in Egypt and the Smithsonian Institution, 1981.

King, David A. "The Exact Sciences in Medieval Islam: Some Remarks on the Present State of Research." Bulletin 4. Tucson, AZ: Middle East Studies Association.

Lunde, Paul, Charis Waddy, and Richard Hobson, and associates. "Science: The Islamic Legacy." *Aramco World* 33(33) (special issue, May–June 1982). New York: Aramco Corporation, 1982.

MAAS Journal of Islamic Science. Published semiannually in January and July by the Muslim Association for the Advancement of Science, Aligarh, India.

#Nasr, Seyyed Hossein. *Islamic Science An Illustrated Study*. London: World of Islam Festival Publishing Co., Ltd, 1976.

Pines, S. "What was Original in Arabic Science?". In *Scientific Change*, edited by A. C. Crombie, pp. 181–205. New York: Basic Books, 1963.

#Qadir, C. A. *Philosophy and Science in the Islamic World*. London and New York: Routledge, 1988.

Rashed, Roshdi, ed., with various scholars. Encyclopedia of the History of Arab Science. 3 vols. London: Routledge, 1996.

Sabra, A. I. "The Appropriation and Subsequent Naturalization of Greek Science in Medieval Islam: A Preliminary Statement." *History of Science* 25 (1987): 223–43.

Sabra, A. I. "The Exact Sciences." In *The Genius of Arab Civilization*, 2d edn., edited by John R. Hayes, pp. 149–69. Cambridge: MIT Press, 1983.

Sabra, A. I. "Islamic Civilization and the Scientific Endeavor." Prepared for catalog (unpublished) of "Heritage of Islam" Exhibition, 1983.

Sabra, A. I. "The Scientific Enterprise." In *Islam and the Arab World*, edited by Bernard Lewis. New York: Alfred A. Knopf, in association with American Heritage Publishing Co., 1976.

Sabra, A. I. "Some Remarks on al-Kindi as a Founder of Arabic Science and Philosophy." In *Dr. Mohammad Abdulhadi Abu Ridah: Festschrift*, edited by Abdullah O. Al-Omar. Kuwait: Kuwait University, Faculty of Arts, 1993.

Young, M. J. L., J. D. Latham, and R. B. Serjeant, eds. *Religion, Learning and Science in the Abbasid Period*. Cambridge History of Arabic Literature series. Cambridge: Cambridge University Press, 1990.

Astronomy

Kennedy, E. S. "Late Planetary Theory." *Isis* 57 (1966): 365–78.

King, David A. *Al-Khwarizmi and New Trends in Mathematical Astronomy in the Ninth Century*. New York: Hagop Kevorkian Center for Near Eastern Studies, New York University, 1983.

King, David A. "Astronomical Timekeeping in Medieval Islam." *Etudes Arabes et Islamiques:* 86—90.
King, David A. *Astronomy in the Service of Islam.* Collected Studies series. Aldershot, UK: Variorum, 1993.
Ragep, F. Jamil, ed. and trans. *Nasir al-Din al-Tusi's Memoir on Astronomy (Al-Tadhkira fi'ilm al-hay'a).* 2 vols. Sources in the History of Mathematics and Physical Sciences. New York: Springer-Verlag, 1993.
Saliba, George. *A History of Arabic Astronomy: Planetary Theories During the Golden Age of Islam.* New York University Studies in Near Eastern Civilization. New York and London: New York University Press, 1994.
Sayili, Aydin. *The Observatory in Islam and its Place in the General History of the Observatory.* North Stratford, NH: Ayer Co., Publishers, 1981.

Astronomical Instruments

Brieux, M. Alain, and associates. *Collection Leonard Linton: Scientific Instruments/Rare Books.* Paris: Alain Brieux, 1980.
King, David A. "A Brief Survey of Islamic Astronomical Instruments." Prepared for catalog (unpublished), "Heritage of Islam" Exhibition, 1983.

Esoteric Sciences

Eliade, Mircea. *The Forge and the Crucible: The Origins and Structures of Alchemy.* Trans. Stephen Corrin. Chicago: University of Chicago Press, 1978.
Savage Smith, Emilie, and Marion B. Smith. *Islamic Geomancy and a Thirteenth Century Divinatory Device.* Studies in Near Eastern Culture and Society, G. E. von Grunebaum Center, University of California. Malibu: Undena Publications, 1980.

Medicine and Pharmacy

Dols, Michael W., trans. *Medieval Islamic Medicine: Ibn Ridwan's Treatise "On the Prevention of Bodily Ills in Egypt."* Arabic text edited by Adil S. Gamal. Berkeley, Los Angeles, and London: University of California Press, 1984.
Hamarneh, Sami K. *Catalogue of Arabic Manuscripts on Medicine and Pharmacy at the British Library.* Cairo: Les Editions Universitaires d'Egypte, in collaboration with the Smithsonian Institution, 1975.
Hamarneh, Sami K. "Islamic Medicine and the Allied Sciences." Prepared for catalog (unpublished), "Heritage of Islam" Exhibition, 1983.
Hamarneh, Sami K. *Health Sciences in Early Islam.* Edited by Munawar A. Anees. San Antonio: Noor Health Foundation, 1985.
Ullmann, Manfred. *Islamic Surveys II: Islamic Medicine.* Edinburgh: University Press, 1978.

Technology

al-Hassan, Ahmad, and Donald R. Hill. *Islamic Technology: An Illustrated History.* Cambridge: Cambridge University Press, 1986.

Hill, Donald R. *Islamic Science and Engineering.* Edinburgh: Edinburgh University Press, 1994.

Islamic Art

Atil, Esin. *Art of the Arab World.* Washington: Smithsonian Institution, 1975.

Sims, Eleanor G. "Painting in Timurid Iran." *Asian Art II* 2 (Spring 1989), 62–80. This article contains a detailed analysis of the sophisticated mathematical elements that can be found in medieval Persian painting.

Islam and the West

Djait, Hichem. *Europe and Islam.* Trans. Peter Heinegg. Berkeley: University of California Press, 1985.

Lewis, Bernard. *The Muslim Discovery of Europe.* New York: W. W. Norton, 1982.

#Sardar, Ziauddin, ed. *The Touch of Midas: Science, Values, and Environment in Islam and the West.* Manchester: Manchester University Press, 1984.

Watt, W. Montgomery. *The Influence of Islam on Medieval Europe.* Islamic Surveys, No. 9. Edinburgh: University of Edinburgh Press, 1972.

Illustration Sources

Figure 1.1. Map prepared by Michael Graham.
Figure 2.1. Reprinted by permission of the Turkish and Islamic Art Museum, Istanbul, MS #1941.
Figure 2.2. Reprinted by permission of the British Library, London, MS Or. 2784, folio 96r.
Figure 3.1. Adapted from a diagram in George Sarton, *A Guide to the History of Science* (Waltham: Chronica Botanica, and New York: Ronald Press, 1952).
Figure 4.1. Reprinted by permission of the British Library, London, MS Or. 1610.
Figure 4.2. Reprinted by permission of the Turkish and Islamic Art Museum, Istanbul, MS 1973.
Figure 4.3. Reprinted by permission of the Bodleian Library, Oxford, Arabic MS. Marsh 139, folio 16.
Figure 5.1. Taken from p. 349, "Unit 1 Aims and Contents of the Four Books," in *The Arithmetic of Al-Uqlidisi* by A. S. Saidan, publ Reidel, 1978, reprinted by permission of Kluwer Academic Publishers, Dordrecht, The Netherlands.
Figure 5.2. Adapted from a diagram in Nawwab, Speers, and Hoye, eds., *Aramco and Its World* (Dhahran: Aramco, 1980).
Figure 5.3. Reprinted by permission of the Rare Book and Manuscript Library, Columbia University, New York, D. E. Smith MS 45(8).
Figure 5.4. Reprinted by permission of the Bibliothèque Nationale, Paris, MS Arabe 2467, folio 84a.
Figure 5.5. Reprinted by permission of Princeton University Libraries, Department of Rare Books and Special Collections, Garrett Collection of Near Eastern Manuscripts, no. B220.
Figure 5.6a. Reprinted by permission of Ray Graham and Associates. Photo by Robert Emmett Bright.
Figure 5.6b. Reprinted by permission of Ray Graham and Associates. Photo by Robert Emmett Bright.
Figure 5.6c. Reprinted by permission of the Toledo Museum of Art, Toledo, Ohio, gift of Florence Scott Libbey, 1912.502.
Figure 6.1. Reprinted by permission of the University Library, Istanbul, MS No. TY 5953.

Figure 6.2. Reprinted by permission of the Library of Congress, Near East Division, Washington, DC, MS 16 (Microfilm #Or. Near Eastern 77).

Figure 6.3. Reprinted by permission of the Spencer Collection, The New York Public Library—Astor, Lenox, and Tilden Foundations, Persian MS 6, folios 89v–90r.

Figure 6.4. Reprinted by permission of the Bibliothèque Nationale, Paris, MS graec 2389, folio 75v.

Figure 6.5a. Diagram prepared by Michael Graham from a diagram in Ake Wallenquist, ed., and Sune Engelbreckston, transl., *Dictionary of Astronomical Terms* (New York: Bantam Doubleday Dell Publishing Group, Inc).

Figure 6.5b. Diagram prepared by Michael Graham from a display prepared by the National Museum of American History, Smithsonian Institution, Washington.

Figure 6.6. Reprinted by permission of the University Library, Istanbul, MS No. F.1418.

Figure 6.7. Photo © Robert Azzi, Woodfin Camp and Associates, Inc., New York.

Figure 6.8. Reprinted by permission of the Musée National, Carthage, Tunisia. Photo by Andrew Myatt.

Figure 6.9a. Reprinted by kind permission of the Trustees of the Chester Beatty Library, Dublin, MS #3673, folios 14r and 15v.

Figure 6.9b. Reprinted by permission of the National Museum, Damascus, A 1727.

Figure 6.10. Reprinted by permission of the Library of Congress, Near East Collection, Prints and Photos Division, Washington, DC, MS 62 (Microfilm #Or. Near Eastern 102).

Figure 6.11. Reprinted by permission of Mr. and Mrs. Arnold Cohen.

Figure 6.12. Reprinted by permission of the University Library, Istanbul, MS #FY 1404, folio 57r.

Figure 6.13. Photo © R. and S. Michaud, Woodfin Camp and Associates, Inc., New York.

Figure 6.14a. Reprinted by permission of Ray Graham and Associates. Photo by Robert Emmett Bright.

Figure 6.14b and 6.14c. Reprinted by permission of Ray Graham and Associates. Photo by Robert Emmett Bright.

Figure 6.15. Reprinted by permission of the University Library, Istanbul, MS FY 1404, folio 56v.

Figures 6.16a and 6.16b. Diagrams originally prepared to accompany J. D. North, "The Astrolabe," *Scientific American* (Jan. 1974): 96–97. © George V. Kelvin/ Scientific American.

Figure 6.17. Reprinted by permission of the Leonard Linton Astrolabe Collection, Pt. Lookout, New York, LL 1152. Photo by Leonard Linton.

Figures 6.18a and 6.18b. Reprinted courtesy of the Adler Planetarium and Astronomy Museum, Chicago, W-89.

ILLUSTRATION SOURCES

Figure 6.19. Reprinted by permission of the Leonard Linton Astrolabe Collection, Pt. Lookout, New York, LL 1304. Photo by Leonard Linton.
Figures 6.20a and 6.20b. Reprinted courtesy of the Benaki Museum, Athens, 13178.
Figure 6.21. Reprinted by permission of the Museum of the History of Science, Oxford. Photo by Corbis-Bettmann.
Figure 6.22. Reprinted by permission of the Leonard Linton Astrolabe Collection, Pt. Lookout, New York, LL 1660. Photo by Leonard Linton.
Figure 6.23. Reprinted by permission of the Leonard Linton Astrolabe Collection, Pt. Lookout, New York, LL 1723. Photo by Leonard Linton.
Figures 6.24a and 6.24b. Reprinted by permission of the Benaki Museum, Athens, 10698.
Figure 6.25. Reprinted by permission of the Adler Planetarium and Astronomy Museum, Chicago, W-94.
Figure 6.26. Reprinted by permission of the Topkapi Library, Istanbul, MS. No. M.1365.
Figure 6.27. Reprinted by permission of National Museum, Damascus. Photo © R. and S. Michaud, Woodfin Camp and Associates, Inc.
Figure 6.28. Reprinted by permission of the National Maritime Museum, London, G7/363-83.
Figure 6.29. Reprinted by permission of the National Museum of American History, Smithsonian Institution, Washington, DC, MM HT 330,781.
Figure 6.30. Reprinted by permission of the Houghton Library, Harvard University, MS Arabic 4285.
Figure 6.31. Reprinted by permission of the Rare Books and Manuscripts Division, The New York Public Library—Astor, Lenox, and Tilden Foundations, folio riii recto.
Figure 7.1. © The Trustees of the National Museums of Scotland, Edinburgh, 1997.
Figure 7.2. Reprinted by permission of the S. K. Hamarneh Collection, Washington, DC.
Figure 7.3. Reprinted by permission of the Department of Oriental Antiquities, British Museum, London, Inv. No. 1888.5-26.1. © The British Museum.
Figure 7.4. Bildarchiv Preussischer Kulturbesitz, Orientabeilung, Staatsbibliothek, Berlin; Landberg 63 i Bl, 52b. © bpk Berlin 1996.
Figure 8.1. Reprinted by permission of the Freer Gallery of Art, Smithsonian Institution, Washington, DC, F48.8.
Figure 8.2. Reprinted by permission of the Bibliothèque Nationale, Paris, MS 5847, folio 119r.
Figure 8.3. Leiden, University Library, Legatum Warnerianum—Oriental Collections, Or. 3101.
Figure 8.4. Courtesy of Spencer Collection, The New York Public Library—Astor, Lenox, and Tilden Foundations, Pers MS No. 9, f.101v–102r.

Figure 8.5. Reprinted by permission of the Bodleian Library, Oxford, MS Pococke 375, folios 3v-4r.
Figure 8.6. Reprinted by permission of the Topkapi Palace Museum Library, Istanbul, R 1633 mük.
Figure 9.1. Reprinted by permission of the Nationalbibliothek, Vienna, A.F. 10, folio 1v.
Figure 9.2a. Photo by W. Denny.
Figure 9.2b. Photo by W. Denny.
Figure 9.2c. Illustration from *Architecture Arabe ou Monuments du Kaire* by Pascal Coste, Paris, 1839; reprinted by permission of the Art and Architecture Collection, Miriam and Ira D. Wallach Division of Art, Prints and Photos, The New York Public Library—Astor, Lenox, and Tilden Foundations, MQT +++.
Figure 9.3. Reprinted by permission of the US National Library of Medicine, Bethesda, Maryland, P 19.
Figure 9.4. Reprinted by permission of the Egyptian National Library, Cairo, MS 100 Tibb Taimur. Photo © R. and S. Michaud, Woodfin Camp and Associates, Inc., New York.
Figure 9.5. © 1996 Bildarchiv Preussischer Kulturbesitz, Orientabeilung, Staatsbibliothek, Berlin; Ms.Or. fol. 91; 130a.
Figure 9.6. Reprinted by permission of the Walters Art Gallery, Baltimore, 52.100.
Figure 9.7. Reprinted by permission of the Millet Library, Istanbul, MS No. T.79. Photo © R. and S. Michaud, Woodfin Camp and Associates, Inc., New York.
Figure 9.8. Reprinted by permission of the Arthur M. Sackler Gallery, Smithsonian Institution, Washington, DC, Vever Collection, S1987.97a-b.
Figure 9.9. Reprinted by permission of the Freer Gallery of Art, Smithsonian Institution, Washington, DC, F32.20v.
Figure 9.10. Reprinted by permission of the Freer Gallery of Art, Smithsonian Institution, Washington, DC, F53.91.
Figure 9.11. Reprinted by permission of the Nationalbibliothek, Vienna, A. F. 10, folio 2v.
Figure 9.12. Reprinted by permission of the Topkapi Palace Museum Library, Istanbul, Arabic MS No. Ahmet III 2127, folio 2 verso.
Figure 9.13. Reprinted by permission of the Osler Library, McGill University, Montreal, MS 7508.
Figure 9.14. Reprinted by permission of the Topkapi Palace Museum Library, Istanbul, MS #2127.
Figure 9.15. Reprinted by permission of the Cora Timken Burnett collection of Persian miniatures and other Persian art objects, the Metropolitan Museum of Art, New York. Bequest of Cora Timken Burnett, 1957, all rights reserved; 57.51.21.
Figure 9.16. Reprinted by permission of the Ashmolean Museum, Oxford, 1978.1683.
Figure 9.17. Reprinted by permission of the University Library, Istanbul, 4689.

ILLUSTRATION SOURCES

Figure 9.18. Reprinted by permission of the Freer Gallery of Art, Smithsonian Institution, Washington, DC, F46.12a-bb.

Figure 10.1. Reprinted by permission of the Minneapolis Institute of Arts, 51.37.37.

Figure 10.2. Reprinted by permission of the Cleveland Museum of Art, CMA 1971.84. © The Cleveland Museum of Art, 1996, Andrew R. and Martha Holden Jennings Fund.

Figure 10.3. Reprinted by permission of the Topkapi Palace Museum Library, Istanbul, MS #H. 2153, folio 109a. Photo © R. and S. Michaud, Woodfin Camp and Associates, Inc., New York.

Figure 10.4. Reprinted by permission of the Pierpont Morgan Library, New York, MS 500, folio 18.

Figure 10.5. Reprinted by permission of the US National Library of Medicine, Bethesda, Maryland, MS P1.

Figure 10.6. Reprinted by permission of the Metropolitan Museum of Art, New York. Anonymous gift, 1956, 56.20.

Figure 10.7a. Reprinted by permission of Ray Graham and Associates. Photo by Robert Emmett Bright.

Figure 10.7b. Reprinted by permission of Ray Graham and Associates. Photo by Robert Emmett Bright.

Figure 10.7c. Reprinted by permission of Ray Graham and Associates. Photo by Robert Emmett Bright.

Figure 10.8. Reprinted by permission of the Bibliothèque Nationale, Paris, MS Arabe 2964, folio 22.

Figure 10.9. Reprinted by permission of Ray Graham and Associates. Photo by Robert Emmett Bright.

Figure 10.10. Photo courtesy of TURESPAÑA, Madrid.

Figure 10.11. Reprinted by permission of the Suleymaniye Library, Istanbul, MS A 3606. Photo © R. and S. Michaud, Woodfin Camp and Associates, New York.

Figures 10.12a and 10.12b. Photos reprinted courtesy *Aramco World*.

Figure 10.13. Reprinted by permission of the Museum of Fine Arts, Boston. Francis Bartlett Donation of 1912 and Picture Fund, 14.5333.

Figure 10.14. Reprinted by permission of the Freer Gallery of Art, Smithsonian Institution, DC, F30.76.

Figure 10.15. Reprinted by permission of the Museum of Fine Arts, Boston, Hervey Wetzel Fund, 22.1.

Figure 10.16. Reprinted by permission of the Freer Gallery of Art, Smithsonian Institution, Washington, DC, F30.73.

Figure 11.1. Adapted by Michael Graham from a diagram translated by Michele de Angelis from Balinus al-Hakim (pseudo-Apollonius of Tyana), in *Sirr al-Khaliqa wa-Sanʿat al-Tabiʿa (Sources and Studies in the History of Arabic-Islamic Sciences, Natural Sciences, Series I)*, Ursula Weisser, ed. Institute for History of Arabic Science: Aleppo, 1979.

Figure 11.2. Reprinted by permission of the US National Library of Medicine, Bethesda, Maryland, A 65.

Figure 12.1a. Reprinted by permission of the Suleymaniye Library, Istanbul, Fatih 3212.

Figure 12.1b. Reprinted by permission of the British Library, London, MS Royal 1267.

Figure 12.2. Reprinted by permission of the Suleymaniye Library, Istanbul, Aya Sofya 2451.

Figure 13.1. Map prepared by Michael Graham.

Figure 14.1a. Reprinted by permission of the US National Library of Medicine, Bethesda, Maryland, WZ 240 A 717 daL 1562.

Figure 14.1b. Reprinted by permission of the US National Library of Medicine, Bethesda, Maryland, WZ 240 fa 975 L 1520 vol IV.

Figure 14.1c. Reprinted by permission of the Boston Medical Library, The Francis A. Countway Library of Medicine, Harvard Medical School, Ballard 603.

Figure 16.1. Map prepared by Michael Graham.

Index

Note: Page numbers in italics refer to illustration captions.

'Abassi, Shafi, *169*
Abbasid dynasty, 30, 45, 61, 65, 132
Abraham (prophet), 77
Adab al-Tabib ("The Physician's Code of Ethics"; al-Ruhawi), 134
Aga Khan Award for Architecture, 225
agriculture, 164, *178*
  irrigation for, *174*
A'Imma, 'Abd al-, 92
Aja'ib al-makhluqat ("The Wonders of Creation"; al-Qazwini), 164, *172*
alchemy, 189–192, *194*
  diagram of cosmology of, *193*
algebra, 47–48, *53*, 205
algorithms, 47
Alhambra Palace (Granada, Spain), *56*, *58*
  gardens of, *179*
Alhazen. *See* Ibn al-Haytham
Al-Jami' fi al-Tibb ("Collection of Simple Diets and Drugs"; Ibn al-Baytar), 139
almanacs (*ruzname*), 82
Al-Qanun ("The Canon"; Ibn Sina), 136, *213*, *215*
Al-Risala al-Shafiya ("The Satisfying Thesis"; al-Tusi), *54*
Anaxagoras, 37
Anaximander, 36

Apollonius, 61, *75*
Arabic (language), 15–17
  translations into, 29–30, 213–216
  translations into Latin from, 209–210
Arabic numbers, 46, *52*
Arabs
  astronomy of, 63
  geography of, 119
  hydraulic technology of, 165
  music of, 49
  pre-Islamic, 27
Archimedes, 165
architecture, 225
'Arifi, *24*
Aristarchus of Samos, 60, 61
Aristotle, 21, 22
  astronomy of, 60, 62
  cosmology of, 37, 39
  Dante on, 218
  influence of, on Muslim philosophers, *25*
  on optics, 196, *198*
  translations of, 210, *214*
  armillary spheres, 67, *87*
art
  of architecture, 225
  Islamic, in eleventh-century, 202
  mathematics in, 56–58

[253]

Artuqid dynasty (Turkey), 166
astrolabes, 66–67, 88–90, 112, 122
   astrolabic quadrants, 99–101
   Moroccan, with date converter, 97
   Persian, 91, 92, 96
   Spanish, 93
   spherical, 95
   universal, 94
astrolabic quadrants, 67, 99, 101
   Egyptian, 98
   North African, brass, 100
astrology, 108–111
   al-Biruni on, 116
   astrolabes for, 88–97, 112
   cosmology and, 36–39
   diagram of cosmology of alchemy and, 193
   geometric device for, 115
   horoscopes for, 114
   observatories for, 65
   Renaissance scientists and, 218
astronomy, 205
   armillary spheres for, 87
   astrolabes for, 88–97, 112
   astrolabic quadrants for, 98–101
   astrology and, 108–111
   celestial globes for, 103, 104
   celestial illustrations of, 70–74
   Muslim instruments for, 65–67, 78
   observational, 64–65
   observatories for, 83–86
   pre-Islamic, 59–62
   of Ptolemy, 74
   theoretical, 68–69
   zij tables for, 79, 81
Athens (Greece), 132
atlases, 120, 121, 126
Audubon, John James, 168
automata, 166, 181, 188
Averroës. *See* Ibn Rushd
Avicenna. *See* Ibn Sina
ʿAyni, Abu Muhammad Mahmud ibn

Ahmad ibn Musa Bada al-Din al-, 70
Ayyubid dynasty, 137
Azhar, al- (school, Cairo), 30

Babylonians
   astronomy of, 59
   cosmology of, 36
   mathematics of, 43–45, 52
Bacon, Francis, 219
Bacon, Roger, 219
Baghdad (Iraq), 201
   library in, 30
   medical specialists in, 132
   observatory in, 65
   scientific manuscripts translated in, 44
   scientists in, 61
Balkhi School, 119
Banu Musa, 46, 165
barber-surgeons, 138, 147
baths, 160
Battani, al-, 65, 210
Battuti, Muhammad b. Ahmad al-, 97
Baytar, Abu Bakr al-, 158
Bedouins, 27
Berbers, 27
Beyazit II Hospital (Edirne, Turkey), 142
birds, 168–169
Birjandi, al-, 106
Biruni, Abu Rayhan al-, 48
   on astrology, 109, 116
   astronomical data collected by, 65
   measurement devices of, 165
   botany, 163, 164, 167
Brahe, Tycho, 218
Brethren of Sincerity (Ikhwan al-Safaʿ), 49, 190–191, 197

Cairo (Egypt), 65, 201
   libraries in, 30
   Qalaoun Hospital in, 143

calendars, 3
  astrolabe to convert dates between, 97
  Islamic, 63, 64
camera obscura, 196, 197, 200
celestial globes, 67, 103, 104
celestial mechanics, 39
chamomile (herb), 154
Charles Martel (king of Franks), 5
Chaucer, Geoffrey, 67
chemistry, 190, 192
China (ancient)
  compass from, 119
  cosmology of, 37
  mathematics of, 44
Christianity
  cultural influence of Islam on, 211
  Protestant Reformation in, 211
  similarities between Islam and, 10–11
  See also Roman Catholic Church
Christians
  cosmology of, 37–38
  Crusades by, 202, 208
  Eastern Roman Empire of, 132
  in Islamic scientific community, 28
  in Islamic world, 7, 13
  Nestorians, 29
  translations from Arabic into Latin by, 209
Columbus, Christopher, 128
commerce, 27, 117–118, 124, 164, 201, 206
compasses, 102, 119
Constantine the African, 210
Copernicus, Nicolaus, 42, 60, 69, 105, 107
Córdoba (Spain), 201
  astronomy studied at, 62
  science academy in, 30–31
cosmology, 36–39
  of alchemy, 193
  of Ibn al-'Arabi, 40
  of Ibn al-Shatir, 42
  of Loqman, 41

Crusades, 202, 208
Ctesiphon (Sasanid empire), 61
Damascus (Syria), 132, 201
Dante Alighieri, 217–218
Darwin, Charles, 227
Delhi (India), 66
De Materia Medica (Dioscorides), 132, 138–139, 149–151, 153, 155, 156, 173
Descartes, René, 218
Dimashqi, Salim Thabit al-, 80
Dioscorides, 132, 138–139, 149–151, 153
Divrigi (Turkey), 142
earth
  calculating circumference of, 48
  geography of, 118–119
  in Greek astronomy, 60, 61
  Muslims' respect for, 163
Eastern Roman Empire, 132
Edirne (Turkey), 142
education
  in sciences, 203
  of women, 31
Egypt (ancient)
  astronomy of, 59
  cosmology of, 36
  hospitals in, 141, 143
  mathematics of, 43
Elements (Euclid), 44
elixirs (*al-iksir;* "philosopher's stone"), 189, 190, 192, 194
encyclopedias, 164, 170–172
Enlightenment, 220
environmentalism, 163
Eratosthenes, 118
ethics
  medical, 134–135, 148
  scientific, 227
Euclid
  Fifth Postulate of, 47, 54
  on geometry, 44, 46
  geometry of, 53

[255]

on optics, 195–197, *198*
translations of, 48, 210
Euclidean geometry, 47
experimental method, 197, 217–219

falasifa, 20
falconry, 164
Faqih Ilyas, Mansur ibn Muhammad ibn al-, 137–138, *144*
Farabi, Abu Nasr al-, 21, 32, 49, *55*, 109
fara'id (division of estates), 205
Farisi, Kamal al-Din al-, 197, 200, 204
Fatimid dynasty (Egypt), 45
Fez (Morocco), *183*, 202
finger reckoning, 45, *51*
Five Pillars, 11

Galen, 131, 132, 137
Kitab al-Tiryaq by, *152*
translations of, *140*, 210
Galileo Galilei, 60, 66, 227
Gama, Vasco da, 120–121
gardens, *177*, *179*
Genghis Khan (khan of Mongols), 7
geography, 117–121
maps for, *125*–*129*
geology, 164
geometry, 46–47
in Islamic art, *56*–*58*
Gerard of Cremona, 210
Ghafiqi, Abu Ja'far al-, *154*
Ghazali, al-, 22, 32, 204
Ghazna (India), 202
God, 22, 39, 226
gold, 189, 190
Gondeshapur (Persia)
hospital at, 133
medical center at, 132
scientific center at, 61
translations at, 28–29, 44
Granada (Spain), *93*, 202
Alhambra Palace in, *56*, *58*, *179*

Greece (ancient)
astrolabes invented in, 89
astronomy of, 59–61, 68
cosmology of, 36–37
geographic knowledge of, 118
mathematics of, 44, 45, 48
medicine and health in, 131, 132
music of, 49
study of optics in, 195–197
Greek (language), 29, 210, 213
Guttenberg, Johann, 210

hadith (Muhammad's writings), 11, 119
on health, 131
learning encouraged by, 17
Hariri, al-, *124*
Harvey, William, 137
Hellenism, 162
Hellenistic thought
on astronomy, 59–61
pursuit of knowledge in, 163
sciences in, 204, 205
See also Greece
Hindus
mathematics of, 44
number system of, 45, *51*
Hipparchus, 61
Hippocrates, 131, 132, 135
translations of, *140*, 210
Hizam, Muhammad ibn Akhi, *158*
horoscopes, *114*
horses, *158*
hospitals, 133–135, *141*–*143*
Hulagu (Mongol ruler), 84
hydraulic technology, 165, 166

Ibn al-'Arabi, 39, 40
Ibn al-Baytar, 139, 204
Ibn al-Haytham (Alhazen)
as astronomer, 48, 62
on mathematics, 47
on optics, *145*, 196–197, *198*–200, 218

[256]

INDEX

ibn'Ali, Sharaf al-Din, *148*
Ibn al-Khatib, 137
Ibn al-Nafis, 137, 204
Ibn al-Shatir, 204
  on astronomy, 42, 69, *105*
  zij by, *8*
ibn Arfa-Ra, 'Ali ibn Musa, *194*
ibn Bakhtishu', Abu Sa'id Ubaydallah, *170*
ibn Baso, Ahmad ibn Husayn, *93*
Ibn Battuta, 120
Ibn Hawqal, 119
ibn Hayyan, Jabir (Geber), 191, 192, *193, 194*
ibn Ishaq, Hunayn, 30, 61
  on anatomy of eyes, 137, *145*
  Greek medical texts translated by, 132, *152*
ibn Khalaf, Ali, *94*
Ibn Khaldun, 32, 109–110, 120, 191
Ibn Qalaoun, Al-Nasir Muhammad, *158*
ibn Qurra, Thabit, 30, 47, 65
Ibn Ridwan, *114*
Ibn Rushd (Averroës), 137
  Dante on, 218
  on existence of God, 22
  persecution of, 205
  translations of, *213, 214*
Ibn Sina (Avicenna), 21–22, 32, 49, *55*, 227
  astrology opposed by, 109
  as geologist, 164
  on optics, 196
  as physician, 136–137, *140*
  translations of, 210, *213, 215*
Ibn Yunus, 65, 79
Ibn Zuhr (Avenzoar), 138
Idrisi, al-, 120, *127*
Ikhwan al-Safa' (Brethren of Sincerity), 49, 190–191, *197*
'ilm al-miqat (timekeeping), 63

India (ancient)
  astronomy of, 61, 66, 68
  mathematics of, 44, 45, 52
  medicine in, 132
  irrigation, 165, *174, 180*
Isfahani, Abu 'l-Faradj al-, 49
Isfahani, Hamid ibn Mahmud al-, *91*
Islam, 1
  alchemy in, 190, *193*
  Arabic language and, 15–17
  astrology and, 109–110
  astronomy and, 62–64
  commerce of, 117–118, *124*
  cultural dialogue between West and, 221
  cultural influence on Christian West of, 211
  current, map of, 228–229
  differences over science within, 203
  eleventh-century decline in science in, 203–205
  expansion of culture of, 8
  expansion of empire of, 5–7, 9
  faith of, 10–15
  fundamentalism in, 225–226
  geography of, 119
  in late-eighteenth century, 208
  learning encouraged in, 17–18
  medicine in, 131–133
  medieval isolation of, 205–206
  music of, 49–50
  philosophy of, 19–23
  post–World War II, 224
  pursuit of knowledge in, 162
  scientific education in, 31
  Scientific Revolution and, 219
  See also Muslims
Islamic calendar, 3, 64
  astrolabe to convert dates in, 97
  timekeeping for, 63
Islamic law, 13–15
  fara'id (division of estates) in, 205

Israel, 224
Istakhri, Abu Ishaq al-, 119, 120, 125, 126
Istanbul (Turkey), 83

Jaipur (India), 66, 85, 86
Jai Singh (emperor, Mughal Empire), 66, 85
Jami, 160
Jazari, Badiʿ al-Zaman al-, 166, 181, 184, 186–188
Jews
  in Islamic scientific community, 28
  in Islamic world, 7, 13
  translations from Arabic into Latin by, 209
jihad (effort), 12
Judaism, 10–11

Kaʾba (Mecca, Saudi Arabia), 63, 77
Kairouan (Tunisia), 202
kalam (theological examination), 19
Kamal, Yusuf, 125
Kemal Atatürk, 223
Kepler, Johannes, 60, 68, 218
Khalil, ʿAli ibn Hassan Muhammad, 92
Khayyam, Omar, 47, 53
Khazini, al-, 164–166
Khvaju Bridge (Isfahan, Iran), 176
Khwarizmi, al-, 47–48, 65, 78, 210
Kindi, al-, 20–21
  on astrology, 116
  on music, 49, 55
  on optics, 195–197
  on reason and faith, 22
kinematics, 62
Kitab al-Manazir ("Book of Optics"; Ibn al-Haytham), 196, 197, 198, 199
Kitab al-Musiqa ("Book of Music"; al-Farabi), 49, 55
Kitab al-Rujari ("The Book of Roger"; al-Idrisi), 120, 127

Kitab al-Tasrif ("Book of Concessions"; al-Zahrawi), 138, 146
Kitab al-Tiryaq ("Book of Antidotes"; Galen), 152

languages
  Arabic, 15–17
  translations, 29–30, 209–210, 213–216
  Latin (language), 29, 209–210, 213
libraries, 30
Loqman, 41, 83
lunar calendar, 64

Maimonides, 137
Majisti, Al- (Almagest; "Great Compilation," Ptolemy), 39, 42, 61, 62, 74
Maʾmun, al-, 49, 65, 132
mandrake (plant), 153
Mansur, al- (caliph), 61
maps, 120–121, 125–129
Maqamat ("Assembly"; al-Hariri), 124
Maragha (Persia), 65, 109
Masudi, al-, 49
Masʿudi, al-, 119–120, 164
mathematics
  algebra, 47–48
  astronomy and, 62
  finger reckoning in, 51
  geometry, 46–47
  in Islamic art, 56–58
  of Khayyam, 53
  music and, 48–50, 55
  numbers for, 52
  number systems in, 45–46
  of optics, 197
  pre-Islamic, 43–45
  trigonometry, 69
Mawsili, al-, 49
measurement technology, 165–166
Mecca (Saudi Arabia), 202

astrolabes to determine direction
  to, 89
  Ka'ba in, 63, 77
qibla (direction of), 78–80
qibla boxes indicating direction to,
  67
Mecca plate, 80
medicine
  devices for measurements in, 186
  great physicians, 135–138, 140
  hospitals and, 133–135, 141–143
  Islamic, 131–133
  pharmacology and, 138–139, 153–157
  pre-Islamic, 131–132
  veterinary, 158
Medina (Saudi Arabia), 202
mental illness, 133
Mesopotamians, 36, 43
Mongols, 65
Muhammad (prophet), 10
  Gabriel and, 24
  on importance of water, 165
  Islamic calendar tied to, 3, 64
  in Islamic theology, 11
  at the Ka'ba, 77
  successors to, 7
  as trader, 117
  See also *hadith*
Murad III (sultan, Turkey), 83
Muruj al-dhahab ("The Meadows of
  Gold and Quarries of Jewels"; al-
  Mas'udi), 120
music, mathematics and, 48–50, 55
Muslims, 1–2
  astrology practiced by, 108–111
  astronomical instruments of, 65–67,
    78, 87–104
  astronomy of, 62–65, 68–69
  commerce of, 117–118, 124
  communal baths of, 160
  cosmology of, 38–39
  current population of, 228

eleventh-century decline in society
  of, 203–205
eleventh-century empire of, 201–202
expansion of Islamic empire by, 5–7
finger reckoning by, 51
hospitals established by, 133–135, 141–143
impact of modern West on, 212
mathematics of, 44–48
music of, 48–50
as naturalists, 163, 167
pharmacology of, 138–139, 153–157
as physicians, 135–138, 140
pilgrimages to the Ka'ba by, 77
technology of, 165–166
travel of, 117–121
Western institutions studied by, 223
See also Islam
Mu'tazila, 20
muwaqqit (timekeepers), 64

natural philosophy, 220
natural sciences, 162–166, 167
  encyclopedias on, 170–172
Neoplatonism, 20
Nestorians, 29
Newton, Sir Isaac, 218
Nizami, 112
Noah (mythical), 122
Normans, 120, 127, 202, 208
numbering systems, 43–46
  finger reckoning as, 51
numbers, 45–46, 52

observatories, 65–66
  astronomical instruments at, 66
  at Jaipur (India), 85, 86
  of Murad III, 83
  at Samarkand, 84, 223–224
  used for astrology, 109
optics, 195–197, 198–200, 218
ornithology, 164

Ottoman Empire, 205, 222
  almanacs (*ruzname*) in, 82
  in eighteenth century, 208
  hospitals in, *141*, *142*
  modernization of, 206
  replaced by Western colonial rule, 223
  in seventeenth century, 211
  Western innovations assimilated by, 212

Palermo (Sicily), 202
Palestinians, 224
Paracelsus, Philippus Aureolus, 218
Paulus of Aegineta, *140*
Persia, 222
  Gondeshapur in, 28–29
  libraries in, 30
pharmacology, 138–139, *153–157*
  botany and, 164
  herbal, 132
pharmacy, 139, *153–157*
  botany and, 164
"philosopher's stone" (elixir; *al-iksir*), 189, 190, 192, *194*
philosophical sciences, 32
physicians, 135–138, *140*
  ethics of, 134–135, *148*
  See also medicine
planets, 61, 68–69, *75*
  Tusi Couple for, *105*, *106*
Plato, 21, 37, 60, 132
prayers
  almanacs for time of, 82
  qibla (direction of Mecca) for, 78–80
  sundials for timing of, 78
  timekeeping needed for, 63, 64
Protestant Reformation, 211
Pseudo-Galen, *178*
Ptolemy
  astrology based on astronomy of, 109
  astronomy of, 38, 39, 42, 61, 62, 68, 71, 74, *105*

geography of, 118, 127
on optics, 195–197, *198*
Tetrabiblos by, 109, *114*
translations of, 210
Pythagoras, 37, 49, *55*, 60

Qalaoun Hospital (Cairo, Egypt), *143*
qanat (underground conduits), 165
Qazwini, Zakariya ibn Muhammad ibn Mahmud Abu Yahya al-, 164, *172*
qibla (direction of Mecca), 78–80
  compasses and, *102*
qibla boxes, 67
quadrants, 67, *101*
  astrolabic, 99
  Egyptian, *98*
  North African, brass, *100*
Qur'an, 10, 11
  Arabic language of, 16
  on astrology, 109
  geography of, 118, 119
  on human responsibility, 228
  interpretations of, 22–23
  Islamic law in, 13–15
  philosophy contained in, 18–19
  on value of scientific knowledge, 31

Rakka (Syria), 65
Raphael, 218
Rashid, al- (caliph), 133
Razi, al- (Rhazes)
  as alchemist, 191, 192, *194*
  medical texts of, 135–136
  on music, 49
  on optics, 196
  as physician, *140*
  translations of, 210, 213, 216
Reformation, 211
Re'is, Piri, 121, *128*
Renaissance, 206, 211, 217–221
Rhazes. *See* Razi, al-

[260]

Roger II (Norman king, Sicily), 120, 127
Roman Catholic Church, 220
  scholasticism of, 217
  scientific inquiry and, 218
  See also Christianity
Ruhawi, al-, 134
ruzname (almanacs), 82

Saladin (sultan, Egypt and Syria), 137
Samarkand (Uzbekistan), observatory at, 65, 84, 109, 223–224
Sarraj, Ahmad Ibn al-, 94, 98
Sarton, George, 35
Sasanid empire, 61
Saudi Arabia, 224, 225
scholasticism, 217
science
  cultural transmission of, 35
  education in, 31
  encyclopedias of, 170–172
  experimental method in, 197, 217–219
  Islamic, eleventh-century decline in, 203–205
  of Islamic empire, 28
  natural, 162–166, 167
  opposition to, 226–228
  pre-Islamic, 26–27
Scientific Revolution, 219
Scotus, Michael, 210
Seljuq Turks, 202, 208
Shariʿa (God's law), 13–15
Shiʿite Muslims, 12
Sicily, 120, 127, 202, 208
Southeast Asia, 206
Spain
  Alhambra Palace (Granada) in, 56, 58
  astronomy in, 62
  Córdoba in, 30–31, 201
  reconquest by Christians of, 202, 208

translations from Arabic into Latin in, 209–210, 213
Sufi, ʿAbd al-Rahman al-, 71
Sufism, 12, 22
Sulayman (the Magnificent; sultan, Turkey), 222
Sumerians, 43
sundials, 67, 78
Sunni Muslims, 12
surgery, 138, 146
  instruments for, 147

Taoists, 37
technology, 165–166
  of armillary spheres, 87
  of astrolabes, 88–97
  of astrolabic quadrants, 98–101
  of astronomical instruments, 65–67
  of celestial globes, 103, 104
  of compasses, 102
  geomantic device for astrology, 115
  hydraulic, 174–182, 188
  of medical devices, 186
  post–World War II, 224–225
  of sundials, 78
  of surgical instruments, 146, 147
  of water clocks, 183–185
telescopes, 66
Tetrabiblos (Ptolemy), 109, 114
theology
  astrology and, 109–110
  cosmology and, 39, 41
  on pursuit of knowledge, 162, 203
timekeepers (muwaqqit), 64
timekeeping (ʿilm al-miqat), 63
  astrolabes for, 66–67, 88–97
  sundials for, 78
  water clocks for, 183–185
Toledo (Spain), 31
toxicology, 152
trade. *See* commerce

trigonometry, 48, 69
  astrolabic quadrants for, *100*
Turkey. *See* Ottoman Empire
Turkics, 65
Tusi, Nasir al-Din al-, 204
  astronomy of, 68, *105, 106*
  Euclidean geometry of, 47, *54*
  trigonometry of, 69
Tusi Couple, 68–69, *105, 106*

Ulugh Beg (ruler of Turkestan), *84, 104, 223*
'Umar, 3
Umayyad dynasty, 30, 132
universities, 27, 33

Valencia (Spain), *180*
veterinary medicine, *158*
vision, 195–197, *198*, 199

water, hydraulic technology for, 165, 166, 174–182, *188*
water clocks, 165, 183–185
Water Court (Valencia, Spain), *180*
water fountains, 177, *179*, *187*
waterwheels, 174–175
women, 31

Yaqut, 120
Yazdi, Muhammad Mahdi al-, *96*
Yazd-Isfahan region (Persia), *96*

Zahrawi, Abu'l-Qasim al- (Abulcasis), 138, *146*
Zarqali, al-, *94*
zero (*sifr*), 45
zij (astronomical tables), 65, *81*
zoology, 163

CPSIA information can be obtained
at www.ICGtesting.com
Printed in the USA
FSHW011840101120
75621FS